PRAISE FOR
MITOCHONDRIA AND THE FUTURE OF MEDICINE

"The topic of mitochondria may seem dry and uninteresting to the uninitiated, but this book makes mitochondria come to life with vivid descriptions accessible even to those with no training in biology. From infertility to aging to cancer and neurological disease, Dr. Lee Know will teach you that mitochondria play a central role in much that we care about in health and disease."

—STEPHANIE SENEFF, PhD, senior research scientist,
MIT Computer Science and Artificial Intelligence Laboratory

"In 1991, when faced with my own health challenges, mitochondria became front and center in my quest for health. Over the last two and a half decades, more and more researchers and clinicians are finding themselves interested in these little powerhouses and proclaiming themselves 'mitochondriacs.' Dr. Lee Know does a brilliant job shedding light on this once ignored organelle and shows us how to care for our most important metabolic system."

—DR. NASHA WINTERS,
coauthor of *The Metabolic Approach to Cancer*

"*Mitochondria and the Future of Medicine* is a tour de force of mitochondria and human health. When it comes to curing chronic disease and extending longevity, it's not an understatement to say that the content of this book will be shaping the future of medicine."

—ARI WHITTEN, best-selling author and
creator of *The Energy Blueprint*

"In *Mitochondria and the Future of Medicine*, Dr. Lee Know takes the mystery out of the evolving science surrounding mitochondria. Here, he clearly and concisely describes mitochondrial structure and function while presenting us with multiple examples of why healthy mitochondria are so crucial to our overall health. Without conjecture or overreach, Dr. Know also provides the rationale behind nutritional strategies that have great potential to improve the status of our mitochondria, a central tenet of healthy aging."

—MIRIAM KALAMIAN, author of *Keto for Cancer*

MITOCHONDRIA
AND THE FUTURE
OF MEDICINE

MITOCHONDRIA AND THE FUTURE OF MEDICINE

The Key to Understanding Disease,
Chronic Illness, Aging, and Life Itself

LEE KNOW, ND

Chelsea Green Publishing
White River Junction, Vermont

Originally self published in Canada by Lee Know, ND,
in 2014 as *Life: The Epic Story of Our Mitochondria*.

Project Manager: Patricia Stone
Developmental Editor: Makenna Goodman
Copy Editor: Deborah Heimann
Proofreader: Nanette Bendyna
Indexer: Linda Hallinger
Designer: Melissa Jacobson

Printed in the United States of America.
First printing February, 2018.
10 9 8 7 6 5 4 3 2   18 19 20 21 22

Our Commitment to Green Publishing
Chelsea Green sees publishing as a tool for cultural change and ecological stewardship. We strive
to align our book manufacturing practices with our editorial mission and to reduce the impact
of our business enterprise in the environment. We print our books and catalogs on chlorine-free
recycled paper, using vegetable-based inks whenever possible. This book might cost slightly more
because it was printed on paper that contains recycled fiber, and we hope you'll agree that it's
worth it. Chelsea Green is a member of the Green Press Initiative (www.greenpressinitiative.org),
a nonprofit coalition of publishers, manufacturers, and authors working to protect the world's
endangered forests and conserve natural resources. *Mitochondria and the Future of Medicine* was
printed on paper supplied by Thomson-Shore that contains 100% postconsumer recycled fiber.

Library of Congress Cataloging-in-Publication Data
Names: Know, Lee, 1976- author.
Title: Mitochondria and the future of medicine : the key to understanding disease, chronic
    illness, aging, and life itself / Lee Know, ND.
Description: White River Junction, Vermont : Chelsea Green Publishing, [2018] | Includes
    bibliographical references and index.
Identifiers: LCCN 2017044901| ISBN 9781603587679 (pbk.) | ISBN 9781603587686 (ebook)
Subjects: LCSH: Mitochondria—Popular works. | Mitochondrial pathology--Popular works.
    | BISAC: MEDICAL / Microbiology. | SCIENCE / Life Sciences / Cytology. | MEDICAL /
    Holistic Medicine.
Classification: LCC QH603.M5 K56 2018 | DDC 571.6/57—dc23
LC record available at https://lccn.loc.gov/2017044901

Chelsea Green Publishing
85 North Main Street, Suite 120
White River Junction, VT 05001
(802) 295-6300
www.chelseagreen.com

Dedicated to H.E.A.L.
"the Knords"

Erin, Aidan, and Hudson

CONTENTS

CHAPTER ONE

The Force

The Origins and Evolution of Mitochondria in Human Physiology

Without the midi-chlorians, life could not exist, and we would have no knowledge of the Force. They continually speak to us, telling us the will of the Force. When you learn to quiet your mind, you'll hear them speaking to you.

—Star Wars: Episode I—*The Phantom Menace*,
Qui-Gon Jinn to Anakin Skywalker

A long time ago, in a galaxy far, far away, there were intelligent microscopic life forms called *midi-chlorians* that lived symbiotically inside the cells of all living things. When present in sufficient numbers, they allowed their symbiotic host to detect the pervasive energy field known as the Force. Midi-chlorian counts were linked to one's potential in the Force, ranging from normal human levels of 2,500 per cell to the much higher levels in a Jedi. The highest known midi-chlorian count (over 20,000 per cell) belonged to Jedi Anakin Skywalker.

Present in all life, midi-chlorians are the same on every world that supports life—in fact, midi-chlorians are necessary for life to exist. In sufficient numbers, midi-chlorians will allow their host organism to detect the Force, and this connection can be strengthened by quieting one's mind, allowing the midi-chlorians to "speak" to their host and communicate the will of the Force.

For many reading this book, I'm sure you're thinking, "What . . . the . . . has he gone completely sideways?" What the heck am I talking about? Well, science fiction fans—and the generation(s) who grew up in the era of *Star Wars*—might have a better chance at guessing that midi-chlorians are a creation of George Lucas . . . or are they?

Midi-chlorians were first conceived by George Lucas as early as 1977. At this time, Lucas sat down with a member of his staff to dictate a number of guidelines for these works, explaining various concepts of his universe. Among them was an explanation of midi-chlorians (even though Lucas did not feel he had the time or opportunity to introduce the concept until 1999, when it was first mentioned during *Star Wars: Episode I—The Phantom Menace*). Explaining why some were sensitive to the Force while others were not was an issue that he needed to address—an issue that he had left unresolved since the original film *Star Wars*.

Midi-chlorians in *Star Wars: Episode 1—The Phantom Menace* are part of a recurring theme throughout the movie—that of symbiotic relationships. What's fascinating to me is that midi-chlorians were loosely based on mitochondria, organelles that provide energy for cells on our non–science fiction, real-world planet. Like midi-chlorians, mitochondria are believed to have once been separate organisms that inhabited living cells and to have since become part of them; even now, mitochondria act in some ways as independent life forms, with their own DNA.

Most readers might remember mitochondria from high school cell biology class, described as the "powerhouses" of the cell—the tiny generators that live inside cells and produce almost all the energy cells need to live. Depending on the type of cell, there are usually hundreds to thousands of mitochondria in each cell. They use the oxygen from the air we breathe to burn up the food we eat to produce useful energy.

Some people might have heard of "Mitochondrial Eve." Since mito-chondria are inherited maternally, if we trace our genetic lineage from child to mother, to maternal grandmother, and so on, Mitochondrial Eve would be the mother of all mothers. (She is thought to have lived in Africa approximately 170,000 years ago. This does not necessarily mean she was the first human; it only means that she is the most recent ancestor common to all humans living today.)

The reason we can trace our ancestry this way is because all mitochondria have their own DNA (the "genes"), which are normally passed on to our children only in the mother's egg, not in the father's sperm. This means that mitochondrial DNA (abbreviated as mtDNA) act like a genetic surname. However, unlike typical Western surnames passed down the paternal line (which can change for any reason, including marriage), mtDNA is fairly constant and unchanging, which allows us to trace our ancestry down the female line. This fact also means that it is usually possible to confirm or disprove familial relationships.

It also makes mtDNA of great use in forensics (to identify people or corpses). One reason why mtDNA is so useful in forensics is that there is a lot of genetic material in each cell. Whereas there are only two copies of the DNA in the nucleus (called *nuclear DNA*, abbreviated as nDNA—the control center of the cell), each mitochondrion contains five to ten copies of its genes. While there is only one nucleus per cell, there are usually several hundred to a couple thousands of mitochondria, meaning there are many thousands of copies of the same mtDNA in each cell.

On the medical side of the story there is the "mitochondrial theory of aging." I'll discuss this in depth ("The Mitochondrial Theory of Aging," page 47), but basically, this theory argues that aging—and many of the diseases that come with it—is caused by a slow degeneration in the quality of mitochondria. This is because during normal cellular respiration—the process where the mitochondria burn up the food we eat using the oxygen we breathe—reactive molecules called *free radicals* are created. These free radicals then go on to inflict damage to adjacent structures, including the DNA in both the mitochondria and nucleus.

Free radicals attack the DNA in each of our cells tens of thousands of times daily. Much of the resulting damage is fixed silently by the extensive repair machinery within the cells, but sometimes these attacks can cause irreversible damage—permanent mutations in the DNA. As the onslaught of free radicals continues day in and day out, these mutations build up over a lifetime. Once the damage reaches a threshold, the cell dies, and slowly over time, tissues start to degenerate with each dying cell. This steady erosion is what's responsible for many age-related degenerative diseases and even the aging process itself.

There are also mitochondrial diseases, some of which might be known by the reader, whether inherited or acquired, that typically affect metabolically active tissues such as the muscles, heart, and brain. This leads to a wide assortment of symptoms depending on the location of the most affected tissues.

The United Kingdom voted in 2015 to legalize a controversial fertility treatment: a technique called *nuclear genome transfer*, a type of mitochondrial replacement therapy. This is where the nucleus is removed from an egg cell (called *oocyte*) of a healthy and fertile female donor (leaving all other components, including the healthy mitochondria), and then the nucleus from the zygote (the fertilized egg) of the infertile woman is transferred into the healthy donor egg. Both ethical and practical concerns have kept this process outlawed throughout the rest of the world, but the United Kingdom continues to push forward, allowing babies to be born with three genetic parents (nDNA from the mother and father, and mtDNA from the donor, or third parent). At the end of 2016, the United Kingdom granted its first license, and the first legal baby using this technique will be born in 2017. (I use the term *legal baby* because this technique was used in 2015 in Mexico, where there were no regulations regarding it, with its three-parent baby born in 2016.)

However, over the last couple decades, one of the most important aspects of the mitochondria has been something that didn't get a lot of media coverage, and that is its role in *apoptosis* (pronounced "A-po-TOE-sis" with the second *p* in its spelling silent), which is programmed cell death or cell suicide. Apoptosis is when individual cells commit suicide for the greater good of the body as a whole.

Previously, apoptosis was thought to be governed by the genes in the nucleus. However, in an eye-opening turn of events starting around the mid-1990s, researchers discovered that apoptosis is actually governed by the mitochondria. The implications for the medical field are profound, especially related to cancer research. Cells are constantly aging or being attacked, resulting in mutations of their DNA. When mutations result in a cell that wants to replicate out of control, it ultimately leads to the dreaded C-word: cancer. Cells failing to commit suicide when directed to do so is now considered the root cause of cancers.

However, the implications run even deeper. Without programmed cell death, complex multicellular organisms might never have had the direction and organization required to evolve in a controlled manner, and the world we know would likely look completely unrecognizable. Sounds confusing, I know. It'll make a lot more sense after I explain it further in "The Evolution of the Eukaryotic Cell," on page 9.

This is in addition to the fact that cells in multicellular organisms (called *eukaryotic* cells) are orders of magnitude larger than single-celled bacteria. There is just no possible way that the energy needs of a eukaryotic cell could be met without mitochondria, as you'll realize shortly.

Although I won't get into the evolution of the two sexes (male and female), mitochondria even help answer the question, "Why do we have two sexes?" Sex between a male and female, while providing intense pleasure for the participants, is actually an inefficient method of procreation. For humans, it requires two parents to produce a single child (most of the time—of course, there are variants here). Clonal reproduction, on the other hand, requires just a mother—the father is not only useless but actually a waste of resources (coincidentally, I was editing this line on Father's Day weekend). Moreover, having two sexes means that only half the population is available to procreate, which is mathematically inefficient. Logically, a better scenario would be if we could procreate with anyone, either because everybody was the same sex or because there was an infinite number of sexes.

But there is a reason why we only have two sexes, and it's mito-chondria, now seen as central to the most widely accepted explana-tion of why: One sex is specialized in passing on its mitochondria to the offspring (eggs from the female), while the other is specialized in making sure its mitochondria are *not* passed on (sperm from the male). I'll elaborate on this when I discuss fertility, infertility, and conception in chapter 2, "Infertility and Mitochondria," on page 113.

Let's Review Some Cell Biology

So now I must forewarn you: This is where it starts to get a little heavy, especially if you don't have a science or biology background. In order to effectively communicate the importance of the mitochondria and the significance of the research contained in this book, I have to discuss a few technical details and ensure all readers have at least a basic understanding of cell biology. Therefore, I feel a quick and dirty review is well worth the few extra pages of reading. If I lose you with the details, don't get tied up in a knot; just try to understand the big picture. Some level of detail is provided, however, so that those with some science background can start to understand the complexity of the picture. So, here we go . . .

The cell is the simplest form of life capable of independent existence, and because of this, it makes up the most basic unit of biology. Single-celled organisms, such as bacteria, are the simplest of cells. They are extremely small, and rarely more than a few thousandths of a millimeter in diameter. Their shapes vary, but are typically either spherical or rodlike in appearance. They are protected from their external environment by a strong, yet permeable, cell wall. Within this wall is the cell membrane—an incredibly thin and delicate, but relatively impermeable, membrane. Bacteria use this membrane to generate their energy. This same membrane is what has become the inner membrane of the mitochondria—arguably the most important membrane in the human body.

Inside the bacterial cell is the *cytoplasm*—the gel-like mass that contains countless biological molecules. Some of these "large" molecules are barely visible through a powerful microscope, even when amplified a millionfold. Among these molecules is the long, coiled structure of DNA—the legendary double helix that was first described by Watson and Crick more than half a century ago. Beyond this, there is not much else we can see. Yet biochemical analysis shows that bacteria, the simplest of life forms, are actually so complex that we still don't know much about their indiscernible organization.

Humans, on the other hand, are composed of different types of cells.* Although cells are considered the simplest basic unit of life, the volume of these types of cells is often a hundred thousand times that of bacteria, and this allows us to see much more inside. There are great structures made of elaborate membranes (called *organelles*) with all sorts of embedded proteins. The organelles are to the cell what organs are to our bodies—discrete entities that are responsible for certain tasks. Also within the cytoplasm are all kinds of large and small vesicles, and a dense network of fibers called the *cytoskeleton*, which gives the cell structural support. Finally, there is the *nucleus*—what most of us would consider the control center of the cell. All these make up the cells that make up the world as we know it, and they are called *eukaryotic cells*. All plants, animals, algae—indeed, essentially every living thing we can see with the naked eye is composed of eukaryotic cells, each one harboring its own nucleus.

Contained within the nucleus, you'll find the DNA. Although the DNA in a eukaryotic cell has the exact same double helix structure as found in bacteria, the way it's organized is very different. In bacteria, the DNA is found as long, twisted loops—known as circular DNA. Don't let its name fool you, however, for it's not "circular" by any means (it looks more like a tangled mess of a ball). The name indicates that there is no beginning or end, just like a circle. There are often numerous copies of this circular DNA in each bacterium, but all are copies of the same genes. In eukaryotic cells, there are usually a number of different chromosomes, which are linear, not circular. Again, this doesn't mean that DNA makes a straight line, but rather that each has two separate and distinct "ends." Also, unlike circular DNA, each chromosome contains different genes. Humans have twenty-three chromosomes, but because we keep two copies of each, we have a total of forty-six. During cell division, these pair up, being

---

* Just to note, here I am not referring to the trillions of bacteria that live in and on the human body, or what we call the microbiome—which the latest research shows is not only incredibly important to our health, but indeed is part of what makes us "human."

joined at the middle, giving them the familiar X-shape we know from science class.

Chromosomes are not just composed of DNA. They are coated with specialized proteins—among them are the *histones*, which not only shield the DNA from harm but also act as gatekeepers to the genes. Histones distinguish eukaryotic chromosomes from that of bacteria, whose DNA is not protected by histones and, therefore, is said to be "naked."

Each of the two strands of the double helix acts as a template for the other. When they are pulled apart during cell division, each strand provides the information needed to reconstruct the full double helix, once again giving two identical copies. The information encoded in DNA is organized into *genes*, which in turn spell out the molecular structure of proteins. Just as all our English words are a sequence of only twenty-six letters, all the genes are a sequence of just four molecular "letters," where the sequence of letters dictates the structure of the protein.

The *genome* (which can be over a billion letters long) is the complete collection of genes in an organism. Each gene (usually thousands of letters) is essentially the code for a single protein. Each protein is a string of subunits called *amino acids*, and it's the precise order of these amino acids that determines the functional properties of that particular protein.

A "mutation" arises when the sequence of letters is changed. This might change the amino acid or the structure of the protein. Thankfully, Nature has built in a certain level of redundancy. Since several different combinations of letters can code for the same amino acid, these mutations don't always result in a structural or functional change in the protein.

This is important because proteins are the fulcrum of life. Their forms and functions are almost limitless, and life as we know it is only made possible because of them. Understanding their function allows us to organize them into several broad categories, such as enzymes, hormones, antibodies, and neurotransmitters.

The whole process of building proteins is controlled by a number of other proteins, with *transcription factors* being the most important.

While DNA contains these genes, it's actually inactive, and transcription factors are what regulate the *expression* of genes. Transcription factors do this by telling the cell to take a particular inactive section of DNA and convert it to an active protein. However, instead of using DNA directly, the cell relies on disposable copies called *RNA*. There are different types of RNA, and each fulfills different tasks. First is messenger RNA (mRNA). Its sequence is an exact copy of the corresponding DNA gene sequence. As its name implies, this messenger passes through the pores in the nuclear membrane, and out into the cytoplasm. From there, it finds one of the many thousands of protein-building factories called *ribosomes*. It is the job of the ribosomes to translate the information encrypted in mRNA into a sequence of amino acids, which make up a particular protein.

Hope you're still with me. As simply as I've tried to describe this, it has taken, and will take, hundreds of scientists their entire careers to tease out the details of an incomprehensibly tiny portion of one sentence in the biology lesson above. This level of understanding, however, should give most readers the ability to understand the significance and inner workings of the mitochondria. Okay, so let's keep going . . .

The Evolution of the Eukaryotic Cell

Although the Greek origin of the word *eukaryotic* means "true nucleus," eukaryotic cells actually contain many other things besides the nucleus—including the mitochondria. Mitochondria were originally independent bacterial entities, which eventually entered other bacterial cells, but instead of becoming digested as they did under typical conditions, they became a *symbiont*, or a partner in a relationship, where both benefit in some way from the presence of the other. You could say that the mitochondria were the original probiotic: a microorganism that provides a benefit to its host.

The theory says that one day, about two billion years ago, one bacterium engulfed another. Initially, each organism was totally autonomous and contained all the genes for independent life. However, after one engulfed the other without disintegrating it, the two experimented with countless different biochemical and genomic arrangements.

This process of trial and error occurred over a mind-boggling expanse of 1.2 billion years, but finally the engulfed bacterium became specialized in energy production (they became the mitochondria) and the rest of this primitive new eukaryotic cell became specialized in structure and function. The acquisition of mitochondria seems to have been *the* decisive moment in the history of life as we know it. If this is true, then the mitochondria deserve all the credit for the abundance of life on Earth as we know it. If not for the mitochondria, the world would not have evolved beyond single-celled bacteria.

Even though mitochondria were once bacteria, the eukaryotic cells they created became vastly different compared to their bacterial origins—and in a number of interesting ways. First, most eukaryotic cells are gigantic when compared to tiny bacteria; most eukaryotic cells have a cell volume about ten thousand to one hundred thousand times that of bacteria.

Second, as mentioned earlier, eukaryotic cells have a nucleus. This is usually a spherical double membrane that houses the compact mass of DNA, which is wrapped in protective proteins. Bacteria, by contrast, lack a nucleus, and their DNA is in a rather primitive and unprotected form.

The third difference is the size of their genomes (the total number of genes). Bacteria typically have far less DNA than eukaryotic cells do. Further, eukaryotic cells have far more noncoding DNA (parts of DNA that don't code for genes) than bacteria do. We previously thought all this noncoding DNA was "junk" DNA—without purpose. Newer research, however, is showing that there is a lot more to DNA than just genes coding for proteins, and that these vast arrays of noncoding DNA (or at least parts of them) actually have numerous functions. Regardless, the huge amount of extra DNA in eukaryotic cells requires much more energy (than bacterial cells require) to copy and to ensure it is copied accurately.

The last major difference I'll discuss is the organization of DNA. As mentioned earlier, bacterial DNA is organized into a single circular chromosome. Although it's anchored to the cell wall, it basically floats freely within the cell. Since bacterial DNA is not covered by a protective protein wrap, it's easily and quickly accessible when needed for replication. Bacterial genes also tend to be organized in functional groups with a similar purpose. Bacteria also carry extra DNA material in the form of tiny rings called *plasmids*. These little rings replicate independently and can be transferred from bacteria to bacteria, relatively quickly. Eukaryotic genes, on the other hand, do not seem to be organized by any identifiable order and the flow is often broken into many short sections with long stretches of noncoding DNA. In order to build a particular protein, many times a large expanse of DNA needs to be read, then spliced apart, before the coding sections are fused together to form an understandable gene that codes for the protein. Further, just getting to these genes is rather complex because the chromosomes are tightly wrapped in the proteins called *histones*. While histones offer a certain degree of protection to the DNA from potential damage, they also block easy access to the genes. When the genes need to be replicated for cell division or to make copies for building proteins, the structure of the histones must be transformed to allow access to the DNA itself. This is the job of another set of proteins, also mentioned earlier, the transcription factors.

Without dragging this discussion on, the most important message I want to get across is that bacteria evolved to be brutally efficient, while most eukaryotic cells are gigantic and incredibly complex beings—and all this complexity comes with an energetic cost.

Many other aspects of the eukaryotic cell also demand great amounts of energy. One example is the internal cytoskeleton of a eukaryotic cell versus the cell wall of a bacterial, or prokaryotic, cell. While they have similar functions—to provide structural support—they are totally different conceptions. The difference is similar to the internal skeleton in a human versus the external shell (or exoskeleton) of an insect or crustacean.

Bacterial walls vary in structure and composition, but in general, they provide rigid exterior support that maintains the shape of the bacterium, preventing it from bursting or collapsing if and when its environment suddenly changes. In contrast, eukaryotic cells usually have a flexible outer membrane, which is given structural stability by the internal cytoskeleton. This cytoskeleton is a highly dynamic structure —constantly being remodeled, which itself requires a significant energy source. This gives eukaryotic cells a huge advantage, as they can change shape and often do so quite vigorously. A classic example of this is a macrophage (a type of white blood cell) engulfing harmful foreign particles, bacteria, or scraps left over from a dead cell.

In fact, virtually every aspect of a eukaryotic cell's life—shape shifting, growing large, building a nucleus, hoarding reams of DNA, multicellularity—requires large amounts of energy and thus depends on the existence of mitochondria. Without mitochondria, higher animals would likely not exist because their cells would only be able to obtain energy from anaerobic respiration (energy production in the absence of oxygen), a process that is much less efficient than the aerobic respiration that occurs in the mitochondria. In fact, mitochondria enable cells to produce fifteen times more energy (as adenosine triphosphate, the cell's energy currency) than they could otherwise—and complex animals, like humans, need large amounts of energy in order to survive.

Mitochondria: They Are the Force

Mitochondria evolved to be the powerhouses, or energy factories, of the cell. They are organelles that act like a cellular digestive system that takes in nutrients, breaks them down, and creates energy for the cell. The process of creating cell energy is known as *cellular respiration*, and most of the chemical reactions involved in cellular respiration happen in the mitochondria.

The mitochondria are very small organelles, yet are shaped perfectly to maximize their hard work. As mentioned, each cell contains hundreds to several thousand mitochondria. The number depends on what the cell needs to do. For example, large numbers are found in heart and skeletal muscle (which require large amounts of energy for mechanical work), in most organs (such as the pancreas, with its biosynthesis of insulin, and the liver, where detoxification takes place), and in the brain (tremendous amounts of energy are required by the nerve cells).

A Spiritual Twist?

The fact that, over billions of years, the formation of the eukaryotic cell happened only once—a chance event—really does make one wonder whether it was guided by a higher power. Indeed, it's possible for science and spirituality (and maybe even some religions) to coexist, as suggested by many academic and philosophical works. Nevertheless, it's important to note that based on the theory of convergent evolution, if we pressed "reset" and started all over again, given enough time (measured in billions of years), many things would happen in a similar way to how they already have. This is because we would run into the same bottlenecks and problems, and with natural selection offering a finite number of ideal solutions to a given adaptation problem, there is a high likelihood solutions would be found in a very familiar way. This idea makes me wonder if life on other planets is biochemically similar based on the convergent evolution theory.

While that's a discussion for an entirely different book, the fact that amino acids (the building blocks of life) have been found in meteorites that are older than our solar system, and that PQQ (a nutrient I'll discuss in chapter 3, "Pyrroloquinoline Quinone [(PQQ)]," page 139) has been found in interstellar dust, it seems that the seeds of life were planted on Earth from the cosmos. We truly are children of the stars.

I suppose this is a bit difficult to accept for some—from an egocentric view, humans seem special; our conscious experience separates us from the mechanical world of physics and chemistry, and perhaps even from the "lower forms" of life. However, it's a fact that the fundamental similarities between all living things outnumber the differences. I write this with full appreciation and acknowledgment of the controversy surrounding the theory of evolution by natural selection and religion. While I prefer to avoid the topic altogether, it's impossible to discuss evolution without acknowledging that massive force of religion pushing back. However, in the face of overwhelming evidence collected over centuries, refuting the reality of evolution is really quite a hollow position to guard. Further, taking such a stance shuts the mind to its amazing story.

Of course there are many unknowns, and many parts of science are speculative, but that is nothing to be ashamed of or dismissed. Certainly, all science is provisional, and we are by no means anywhere near the point of knowing all. But when observations of Nature contradict a theory—no matter how revered, ancient, or popular—the theory will be unceremoniously abandoned, and the hunt for a new and more accurate theory will be intensified. In fact, this is how we ended up with our current understanding of mitochondria: Many theories have been proposed, challenged, tested, and either made stronger or abandoned. This is what science is: a constantly changing base of knowledge.

For traditional religions, it's important to evolve themselves and develop a new paradigm where evolution is woven into the teachings, where, as I alluded to, that evolution was guided by a higher power. However, for hard-core scientists, it's equally important to recognize that even though we think we know lots, we really don't know much at all. It's humbling to remember that everything we know regarding the known universe and our reality, from the most fundamental chemistry to incomprehensibly complex quantum physics— everything—is estimated to be only about 4 percent (at least according to Neil deGrasse Tyson). We don't know about, let alone understand, 96 percent of our universe and our current reality. Thinking we truly know up from down puts us at the same level as those who believed the world was flat.

Is what's written in this book the final word? Likely not. Just about everything else we thought to be true in the past has turned out to be completely false or not entirely accurate, but this is the state of our knowledge at this point in time, and I eagerly await more evidence to strengthen our current knowledge base or send us on a completely (or just slightly) different path.

Any form of life that can't generate its own energy is essentially dead. There is no life without energy. Breathing supplies the blood with oxygen, which in turn gets transported to just about every one of the trillion cells in the body. The cell delivers this oxygen to the mitochondria, where it is used to turn glucose, fatty acids, and sometimes amino acids into energy via cellular respiration, or more specifically *aerobic respiration* (producing energy by the use of oxygen). Although it's hard to believe, gram for gram, we are likely the most powerful energy producers in the universe. In fact, in an interesting calculation outlined by Nick Lane in his book *Power, Sex, Suicide*, it seems we produce ten thousand times more energy (per gram) than the sun every second.

This calls for another quick aside on popular culture, this time the *Matrix*, not *Star Wars*. In this film, the Machines meet their energy needs by harvesting the energy produced in vast "human energy farms." From Lane's calculations, this would make sense. He also points out that some energetic bacteria, such as *Azotobacter*, outperform the sun by a factor of fifty million. This begs the question, "Why hasn't someone taken *Azotobacter* and created a source of clean, organic energy?" I can't be the only one with these trillion-dollar ideas!

The Basics of Mitochondria

Most illustrations depict mitochondria as rodlike, even though they are able to take many different shapes. They are, indeed, quite flexible,

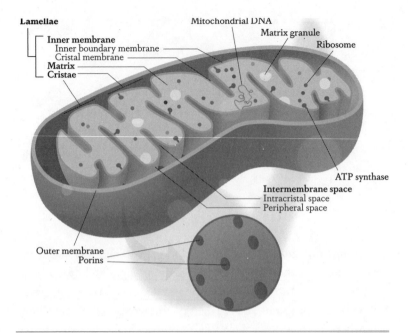

Lamellae

Mitochondrial DNA

Matrix granule

Ribosome

Inner membrane
Inner boundary membrane
Cristal membrane
Matrix
Cristae

ATP synthase
Intermembrane space
Intracristal space
Peripheral space

Outer membrane
Porins

FIGURE 1.1. The structure of a single mitochondrion depicting the double-membrane architecture, with the inner membrane folded over many times, increasing its surface area.

can divide in two like bacteria, or fuse together to form complex structures. Studies show they don't lie still, but constantly move around to the areas they are needed. Their movements appear to be linked to the network of microtubules (think of these as the bones of the cell, the "cytoskeleton" that gives shape to the cell) and are likely transported along the network by motor proteins.

Metabolically active cells, such as those in the heart, muscles, and brain, have thousands of mitochondria. The egg cell (oocyte) has a whopping one hundred thousand mitochondria. In contrast, sperm usually have fewer than one hundred. Red blood cells and skin cells have very few, if any at all. By weight, up to 10 percent of the human body is mitochondria. In numbers, there are about ten million billion. The saying "power in numbers" seems rather fitting.

They were once bacteria, and their appearance and size still resemble those of bacteria. However, unlike bacteria, mitochondria are separated from the rest of the cell by a smooth and continuous outer membrane (instead of a cell wall). Their inner membrane is similar to the bacterial membrane, but convoluted into folds called *cristae* (see figure 1.1).

The cristae vastly increase the inner membrane's surface area inside the organelle. Since this membrane represents the principal site of energy production, the structure of cristae maximizes the area where energy can be produced. On this membrane, energy is produced by transferring electrons down a chain of molecules. This respiratory chain, descriptively known as the "electron transport chain" (ETC), and the various enzymes responsible for energy synthesis are all located in and on the inner membrane.

The space inside (the matrix—no, not the film, but the innermost part of the mitochondria) contains the enzymes of the tricarboxylic acid (TCA) cycle, also known as the Krebs cycle, or citric acid cycle. The resulting molecules produced by the TCA cycle (NADH and $FADH_2$) are fed into the ETC, and the two enzyme systems are located in close proximity to each other so that it can all happen smoothly, without lag.

The Basics of Cellular Respiration and Oxidative Phosphorylation

Of course, every child knows we've got to breathe and eat to live, but the real question is *why*? Why (or how) does supplying our body with oxygen and fuel endow us with this life-giving energy? Cellular respiration is the most important job the mitochondria have. The enzymes of the TCA cycle and the ETC take the molecules resulting from the breakdown of food and combine them with oxygen (O_2), and this results in the production of energy. The mitochondria are the only places in the cell where oxygen can be combined with the food molecules to keep the cell full of energy.

While this explanation would be sufficient for most, we need to understand this to a much greater depth to appreciate the implications related to health and disease, which after all is likely why you're reading this book.

Let's begin with the initial stages of glucose metabolism, called *glycolysis*, which occurs in the cytosol. This is where glucose is converted to a compound called *pyruvate* through a series of chemical reactions. Pyruvate is then transported into the mitochondrial matrix, where another set of reactions convert it to acetyl-CoA. Then the real magic can begin, because acetyl-CoA is the start of the TCA cycle, where the final extraction of energy from food is optimized, yielding carbon dioxide (CO_2, which we will exhale) and two types of energy molecules: NADH and $FADH_2$. Similarly, the breakdown of fatty acids also yields acetyl-CoA, which again goes through the TCA cycle.

The next phase is called *oxidative phosphorylation*, which takes place in the inner mitochondrial membrane. The high-energy electrons from NADH and $FADH_2$ are transferred through a series of carriers in the ETC, and ultimately react with oxygen to yield water. With each step in the ETC, the energy released from these electron transfer reactions is used to pump protons (hydrogen atoms) from

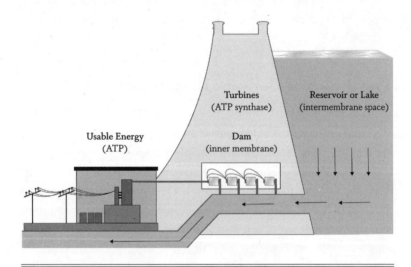

FIGURE 1.2. The process of creating energy in the mitochondria follows the same general principles as a hydroelectric dam. As water (protons) fills a reservoir (intermembrane space) that is contained by a dam (inner membrane), pressure builds up. This pressure forces water to flow out through a channel in the dam, which is used to drive turbines to create hydroelectric electricity.

the matrix to the intermembrane space. This creates a high concentration of protons between the membranes and a low concentration in the matrix. This concentration difference (called *gradient*) is stored potential energy. The high concentration of protons in the intermembrane space wants to flow "downstream" into the matrix and does so through specialized channels, which then create adenosine triphosphate (ATP), the universal energy molecule that all cells use to do their work. Think of this as pumping water (protons) into a reservoir (intermembrane space) contained by a dam (inner membrane). As water flows out through a channel in the dam, it is used to drive turbines to create hydroelectric electricity (see figure 1.2).

This is a very efficient process for taking the energy stored in food to make ATP. Basically, all the absolutely essential things our bodies do to stay alive (e.g., breathing and eating) are to provide the mitochondria with the compounds they need to produce energy. You could say—from a depressing, reductionistic view—that we live only to support our mitochondria.

A Game of Hot Potato: The Electron Transport Chain (ETC)

Four membrane-bound complexes have been identified in the mitochondria; three of them are what we call *proton pumps*. Each is an extremely complex structure that is embedded in the inner membrane. If you refer to figure 1.3, you'll see the various components of the ETC. By following the flow of electrons (e^-) down the ETC, you can see where the protons (H^+) are pumped. Complex I accepts electrons from NADH, and passes them to coenzyme Q10 (CoQ10, denoted in the figure as Q). CoQ10 also receives electrons from Complex II. From there CoQ10 passes electrons to Complex III, which passes them to cytochrome c. Cytochrome c passes electrons to Complex IV, which takes the electrons and two hydrogen ions (H^+) and reacts them with oxygen (O) to form water (H_2O).

However, it's critical to understand that the passage of electrons down this chain is not always 100 percent efficient. A small percentage of electrons are fumbled in this molecular game of hot potato, and

Carbon Monoxide Poisoning

In carbon monoxide poisoning, this toxin displaces oxygen as the final acceptor of the electrons coming down the ETC. The final destination for electrons is no longer available, and when this happens, cellular respiration stops because the electrons no longer have their exit. Unless the carbon monoxide is removed, the mitochondria will die—causing the cells to die, and ultimately killing the person who has been exposed to carbon monoxide.

these leak into the matrix. The rogue electrons then prematurely react with oxygen, resulting in the formation of *superoxide*—a potentially dangerous free radical. Free radicals are highly reactive molecules that contribute to "oxidative stress." This process has been implicated in a number of diseases and even aging itself, as I'll discuss shortly.

In fact, those familiar with the concept of free radicals might be interested to know that the ETC is *the* main site for endogenous free-radical production (free radicals produced within us, as opposed to ones generated by other sources, like environmental pollutants). All this will make more sense in a bit. For now, let's finish discussing the ETC and its components.

Complex I: The First Step of the ETC

Also known as *NADH dehydrogenase*, Complex I is a large molecule made of forty-six protein subunits. It removes two electrons from NADH and transfers them to a lipid-soluble carrier, ubiquinone (oxidized CoQ10, or simply "Q"). In a two-step process, this "reduces" the CoQ10 to ubiquinol (QH_2), and pumps four protons (H^+) across the membrane, creating a proton gradient. This is *the* primary site within the ETC where electrons leak to produce harmful superoxide free radicals.

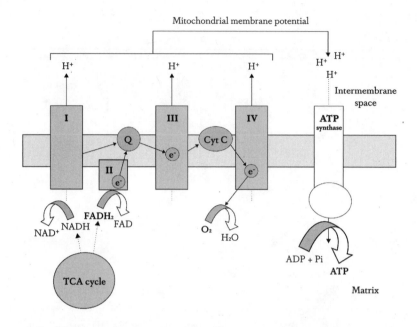

FIGURE 1.3. The electron transport chain (ETC), including ATP synthase. The TCA cycle produces NADH and FADH₂, which enter the ETC at Complex I and Complex II, respectively. Both complexes pass the resulting electron (e^-) to coenzyme Q10 (Q), and the electrons continue until ultimately reacting with oxygen (O_2) to create water (H_2O). Protons (H^+) are pumped at Complexes I, III, and IV, creating the proton gradient. Protons flow back through ATP synthase to create ATP.

Complex II: The Second Step, and a Shortcut to the ETC

This unique complex, also known as *succinate dehydrogenase*, is directly involved in both the TCA cycle and ETC. It's a small complex, consisting of only four protein subunits and the only complex in the ETC that does not pump protons. Its purpose is to deliver additional electrons from succinate to CoQ10 (via FADH₂). Other electron donors (such as fatty acids) also enter the ETC at Complex II via FADH₂.

Complex III: The Twins That Are Master Jugglers of Electrons

Complex III, also known as *cytochrome bc1 complex*, is actually a *dimer*, which basically means that it consists of two identical simpler

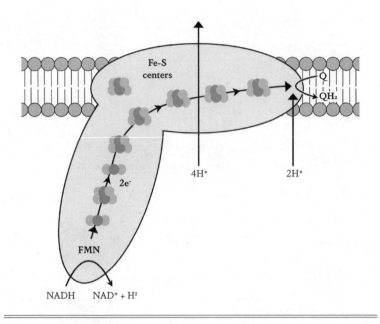

Figure 1.4. Complex I accepts electrons from NADH, passes them through iron-sulphur (Fe-S) clusters to coenzyme Q10 (Q). This results in four protons (H⁺) being pumped from the matrix into the intermembrane space.

complexes. Each part of the dimer comprises of eleven protein subunits, giving twenty-two in total.

This is where the Q cycle occurs, which is a multistep process whereby ubiquinol (reduced CoQ10) is converted to ubiquinone (oxidized CoQ10). In the process, a net of four protons is pumped to contribute to the proton gradient.

This is the second most prevalent site in the ETC where electrons can fall out and react with oxygen to form superoxide free radicals.

Complex IV: The Creator of Water

Complex IV, also called *cytochrome c oxidase*, is made up of thirteen protein subunits. There, four electrons are removed from four molecules of cytochrome c and transferred to molecular oxygen (O_2), producing two molecules of water. This results in four protons being pumped across the membrane, contributing to the proton gradient.

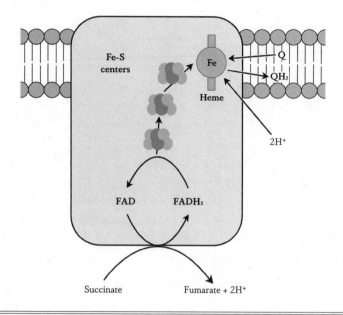

FIGURE 1.5. Complex II is also a part of the TCA cycle, where FADH$_2$ is produced. Electrons from FADH2 are then passed through iron-sulphur (Fe-S) clusters to coenzyme Q10 (Q). This is the only complex that doesn't pump protons.

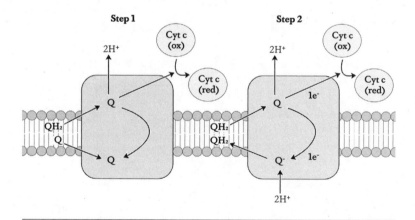

FIGURE 1.6. Complex III accepts electrons from the reduced form of coenzyme Q10 (QH$_2$) in a multistep process called the *Q cycle*. Electrons make their way to cytochrome c (Cyt c), and four protons (H$^+$) are translocated to the intermembrane space.

How Camels Store
Their Water in the Desert

This is a good time to discuss that camels don't actually store water in their humps, as commonly believed by children. Their humps are large deposits of fat. This fat not only serves as a large store of energy, but when this fat is metabolized through oxidative phosphorylation, water is generated at Complex IV as just described (approximately 1 gram—or 1 milliliter—of water for every 1 gram of fat burned). This is partially why camels can go so long without drinking water (along with other adaptations).

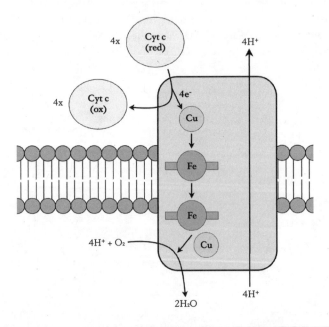

FIGURE 1.7. Complex IV accepts electrons from cytochrome c (cyt c), pumps four protons into the intermembrane space, and passes the electrons to the terminal/ultimate receiver, oxygen (O_2), to create harmless water (H_2O).

Cyanide Poisoning and Suicide

Cyanide, the familiar poison (implicated in the mass suicide in Jonestown, Guyana, and historically given to certain military personnel to ingest in the event of capture), kills by shutting down the ETC. Specifically, it inhibits the activity of Complex IV by binding to the iron (Fe) component, which stops the flow of electrons. The newest approved antidote (at least in the United States) is *hydroxocobalamin* (a form of vitamin B_{12}), which reacts with the cyanide to form *cyanocobalamin* (a form of B_{12} that can be found in most nutritional supplements), which can then be safely eliminated by the kidneys.

Supercomplexes: Optimizing the Speed of Electron Flow

I should point out that what I just described is how we were taught about the ETC in high school and university biology courses. In addition to the four complexes described previously, there is ATP synthase (discussed next, and what some have referred to as Complex V). So in total, we can think of the mitochondrial ETC chain consisting of five enzyme complexes that are responsible for ATP generation. However, this paradigm of the ETC—as discrete enzymes diffused in the inner mitochondrial membrane—has been recently replaced by the "solid state supercomplex" model wherein the respiratory complexes associate with one another to form supramolecular complexes. This arrangement allows for highly efficient electron transfer with the distance an electron must travel between complexes being reduced to several nanometers.

To make matters even more complicated, not only is the existence of supercomplexes debated, but it looks like there could be varying formations of these supercomplexes—for example, the one called *respirasome* includes Complexes I, III, and IV. However, there are supercomplexes of just Complexes I and III, and Complexes III and

IV. These associations will also dictate the pool of CoQ10 and cytochrome c available to these supercomplexes.

Also, there is evidence that certain health conditions are linked to disassociation of the components of these supercomplexes, and while I won't talk specifically about the health conditions this has been linked to (this model is new, and therefore it hasn't been investigated enough), I mention it quickly as a way to illustrate how our knowledge on this is constantly evolving and expanding.

ATP Synthase: Coupling the ETC with Oxidative Phosphorylation

ATP synthase, also known as *ATPase* or *Complex V*, is an important enzyme, as it is the final step in the long chain of events that culminate in the synthesis of ATP. This enzyme is what connects the proton gradient (created by the ETC made possible by the presence of oxygen) to *phosphorylation*—the process of adding a phosphate to adenosine diphosphate (ADP), which creates ATP—all in all, known as *oxidative phosphorylation*.

This large enzyme is the smallest known machine. There are some really cool animations on the internet that show it visually, and I encourage you to check them out when you have some time. This rotary motor, constructed from many tiny moving protein parts, has two main components: the drive shaft that is inserted straight through the membrane from one side to the other, and a very large rotating head that is attached to the drive shaft. The high concentration of protons on the outside of the membrane wants to flow downstream and does so by passing through the drive shaft to rotate the head. In humans, a full rotation of the head requires ten protons and releases three molecules of ATP.

The use of proton pumps to store potential energy in the form of an electrochemical gradient, and then harnessing that energy as it comes across a membrane to create chemical energy, might seem like a strange way to create energy. Yet this seems to be similar across all forms of life on Earth.

Photosynthesis in plants happens in a similar way. However, in that case, the sun's energy is used to pump protons across the membrane in *chloroplasts* (what the mitochondria have evolved into in plant form). Bacteria, being the ancestor to the mitochondria, also function in the same way—by generating a proton gradient across their cell membrane and loosely contained with the help of their cell wall. However, in contrast to humans and mammals, the electrons in plants pass down the ETC to a terminal electron acceptor that can be one of many different molecules, not just oxygen. Nevertheless, in every case, the energy extracted from the ETC is used to move protons across a membrane. This concept is so universal, it seems that pumping protons across a membrane is a central signature of life on Earth.

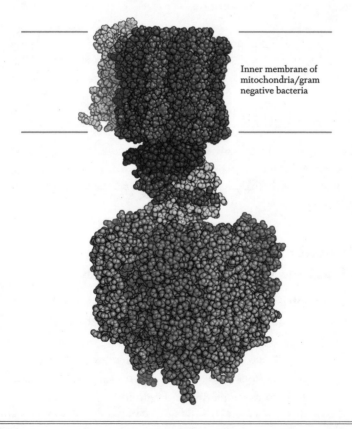

Inner membrane of
mitochondria/gram
negative bacteria

FIGURE 1.8. Molecular representation of ATP synthase showing its orientation and complexity.

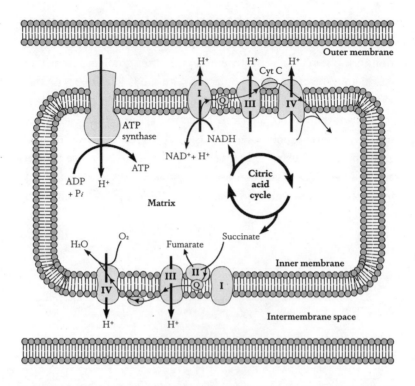

FIGURE 1.9. Illustrative summary of energy production via oxidative phosphorylation, both through Complex I (*top half*) and through Complex II (*bottom half*).

Mitochondrial DNA: A Curious Relic of Ancient History

After the initial bacterial merger that created the eukaryotic cell, the engulfed parasitic bacterium (that eventually went on to form the mitochondria) lived in the lap of luxury. The host bacterium provided many of what it needed to survive, and therefore, it got lazy, so to speak. Many functions were carried out by the host, so there was little need to keep its lumbering and redundant DNA. Why would it need to code for proteins that the host's DNA also coded for? Nature thought the same—and because Nature is relentlessly efficient, the parasitic bacteria began to shed the excessive and redundant genes.

This is no big deal if the genes that are lost are not critical ones. On the flip side, however, cells will surely die if they lose essential genes. The example Lane uses in his book is how our primate ancestors lost the gene for making vitamin C a few million years ago. Thankfully, their diet was rich in fruit, which is a great source of vitamin C. So losing this gene was not catastrophic, and they survived and flourished. How do we know we once had this gene, but then lost it? Most of the gene is still there in our "junk" DNA, and the remaining sequence parallels the functional gene for vitamin C production in other species.

The loss of genes in the name of efficiency is quite common, and in fact, bacteria can lose unnecessary genes in a matter of hours or days. Why is this efficient? Cell division, which is how bacteria reproduce, is highly energy-intensive, and bacteria (relative to eukaryotic cells) produce very little energy. The smaller bacteria's DNA is, the less energy it needs to copy all that DNA for its daughter cells. The efficiency of this gene loss on the main bacterial chromosome is illustrated by their low amount of "junk" DNA.

You might be thinking this is *not* efficient, because the bacteria could end up needing their lost genes down the road. However, throwing away genes is not as reckless as it might first seem. Why? Bacteria can also pick up the same genes again, as well as others, when they need to at a later point, when a concept called *lateral gene transfer* comes into play. A bacterium will pick up DNA from its surroundings (from dead cells or other bacteria) by a form of copulation known as *bacterial conjugation*—not so different from the "transfer" of DNA when humans copulate. (Well, okay, maybe it's a little different, but my point is that bacteria can—and do—accumulate new genes.) The fact that bacteria acquire genes through lateral gene transfer is why I think Big-Agri and the biotech industries need to do far more research before genetically engineered crops and animals are entered into the food chain. The engineered genes can be picked up by the bacteria in our gut or the bacteria in our livestock's gut, and there are countless other ways those genes could "escape" and wreak irreversible havoc throughout the animal and plant kingdoms.

All this gaining and losing genes keeps things in a constant state of flux, which is a good thing, because at any given moment in time, it's unlikely that all redundant genes will be lost from a given community of bacteria. At least a fraction of the bacteria in a particular community is likely to have retained a fully functional copy of that redundant gene. When the surrounding environmental conditions change and the need for that gene becomes important, those bacteria with the gene can pass them around to the other bacteria by lateral gene transfer. Such willingness to share genes explains how things like antibiotic resistance can spread so quickly throughout an entire population of bacteria, and the reason why regulatory bodies, such as Health Canada and the US Food and Drug Administration, require proof that probiotic strains used in supplements do not display antibiotic resistance (because those genes can easily be passed to potential pathogenic bacteria in the gut).

The speed of lateral gene transfer is much faster with the tiny chromosome rings called *plasmids* (which I discussed earlier), but bacteria can also transfer genes that are part of their main chromosomes—it just happens more slowly. Any gene that is not used regularly or not needed at that point in time will tend to be cast aside in favor of more rapid and efficient replication.

As descendants of bacteria, mitochondria lost most of their genes through this process. But if the mitochondrion was a spoon-fed parasite living in its host, why did it need to retain any genes at all? That's a good question—especially when considering that each cell has several hundreds to thousands of mitochondria, and each mitochondrion has five to ten copies of its DNA—and Lane offers an in-depth discussion of this question. The amount of mitochondrial DNA in each cell is a heavy burden when it comes time to divide: When the mitochondria divide, called *mitochondrial fission* or *mitochondrial biogenesis*, or when the cell divides, all those mitochondrial genes must be copied. Further, every mitochondrion must maintain its own genetic translation apparatus and protein-building ribosomes. This process doesn't seem efficient for descendants of bacteria, which thrive on uncompromising efficiency.

Further, there are potentially catastrophic consequences when mitochondria with different genomes are present within the same cell (for example, when paternal mitochondria from sperm survive and coexist with the maternal mitochondria from the egg, which usually results in termination of the pregnancy). This could be avoided altogether if all the mitochondrial genes had been transferred to—and contained within—the nucleus.

Another liability is that the exposed and defenseless genetic material in the mitochondria is located right next to the ETCs, from which destructive free radicals are generated and released. These free radicals can damage the mtDNA and cause mutations, potentially resulting in the demise of the mitochondria (and this can result in numerous health conditions, including cancers, as I'll discuss later).

So, why didn't all the mitochondrial genes get transferred to the nucleus? The fact that they still remain in the mitochondria (despite nearly two billion years of evolution and many good—and logical—reasons why they *should* have been transferred to the nucleus) gives us a clear indication that there must be a reason—and it must be a convincingly fantastic reason.

The reason for retaining some genes seems to be tied to the control over oxidative phosphorylation. The speed of oxidative phosphorylation is very sensitive to the constantly changing situation in energy demands, often fluctuating from minute to minute (depending on whether we're awake, sleeping, exercising, sedentary, fighting an infection, writing an exam, digesting food, reading this book, and so on). These rapidly changing scenarios require the mitochondria to adapt their energy production at a cellular level, and each cell (depending on whether it's a muscle cell, brain cell, white blood cell, intestinal cell, liver cell, and so on) needs to respond individually.

In order to respond efficiently to these swift changes in energy demand, the mitochondria need to maintain a certain level of on-site control, and this control comes in the form of the mtDNA retaining certain genes. The reactions that take place in the ETCs in the inner mitochondrial membrane must be tightly regulated locally at the level of each individual mitochondrion. It would not be efficient, nor would the cell be able to respond quickly to abrupt changes in energy demand, if the entire process were controlled at a distance by the genes tucked far away in the nucleus.

Make sense so far? Good. Now let's talk about supply and demand before I go deeper into answering the question of why mitochondria retained some genes. Remember, the entire process from the individual complexes in the ETC to ATP production by ATPase is coupled like the gears in a machine, where the speed of one gear controls the speed of the next. When energy demand is high, electrons flow down the ETC rapidly, the protons are pumped swiftly, and the proton gradient rises quickly (the proton reservoir "fills up"). The greater the proton gradient, the greater the pressure to form ATP quickly as protons are forced through the ATPase.

However, in a situation where there is no demand for ATP, oxidative phosphorylation will continue to turn all the ADP and phosphate into ATP. Because the cell is not using the ATP (which would convert it back to ADP and phosphate), the ATPase is forced to shut down due to lack of raw materials. When this shutdown happens, protons can no longer pass through the ATPase, and the proton reservoir fills up. When the proton gradient is too high, the small amount of energy released as electrons flow through the ETC is no longer sufficient to pump the protons against this strong gradient. This lack of proton pumping causes the electron flow down the chain to slow and stop. Don't worry, though, things pick up again once the energy demand increases and the cell uses up some ATP (resulting in more ADP and phosphate raw material for the ATPase). This slowdown is why exercise—using up the ATP—is so important; I'll discuss exercise later, too.

There could also be a shortage of oxygen supply—for example, during a stroke when the blood flow to the brain cells is cut off. When this oxygen shortage happens, there is nothing to remove the electrons at the end of the ETC, and therefore, the electrons back up—a traffic jam, so to speak—and oxidative phosphorylation stops. In any of these situations, when electrons back up, they can leak and create free radicals.

However, in addition to supply and demand, we also need to consider the various components of the ETC. Each component can either be reduced (they have an electron) or oxidized (they don't have an electron), but they can't be both at the same time. For example, if Complex I already has an electron, it can't take on another one until it has passed on the first to the next carrier in the chain, ubiquinone. The ETC will be held up until it has passed on this electron. Similarly, if Complex I doesn't have an electron, it can't pass on anything to ubiquinone until it has received an electron (from NADH). The ETC will be held up until it receives one.

This *reduction-oxidation*, or "redox" as it's more commonly known, is an important emerging field of medical study. There are thousands of ETCs in each mitochondrion (approximately ten thousand ETCs per mitochondrion—an astonishing number), and oxidative phosphorylation will proceed most efficiently when there is a balanced 50:50 ratio between oxidized and reduced electron carriers.

However, losing this balance not only slows down oxidative phosphorylation and energy production, but it can also wreak havoc in the mitochondria. This is because each electron carrier in the chain is reactive. If electron flow is progressing normally, each carrier will likely pass on its electrons to the next carrier, which has a slightly greater desire for that electron than did its predecessor. However, because carriers cannot be both oxidized and reduced at the same time, if the next carrier already has an electron, then the ETC becomes blocked, and there is a chance that the electron will be prematurely transferred to oxygen instead. When oxygen receives an electron from any carrier other than Complex IV (the last carrier in the ETC), it forms the toxic free radical superoxide. This is not necessarily a bad thing (as I'll discuss later), but for now let's just consider that superoxide typically goes on to inflict damage to all sorts of biological molecules. We normally don't want this. It's just like trains traveling down a railroad: If a train doesn't leave a particular station in time, the next train can't arrive. In the best-case scenario, this causes a backup. However, if an approaching train didn't get the message that the station is occupied, and doesn't hit the brakes soon enough, there is a good chance that a crash will result and the train's cars will derail, causing all sorts of damage.

So maintaining the 50/50 redox balance allows both a speedy and efficient electron flow down the ETC, but also minimizes the risk of superoxide free-radical formation. However, maintaining this balance also depends on the number of electron carriers within the ETC relative to one another. For example, if a mitochondrion has an abundance of Complex I that have accepted an electron from

NADH, but not enough ubiquinone, many of those "full" Complex I entities will fumble their electron, only to be picked up by oxygen. And just like everything else in the body, the relative number of each component of the ETC is in a state of constant flux—constantly being degraded and replaced.

A Radical Signal:
The Positive Side of Free Radicals

Now, after this roundabout detour of a discussion, we can progress to the next step in answering the original question: Why do mitochondria need any genes at all? Let's consider a hypothetical scenario and imagine that there are one thousand mitochondria in a particular cell, each with about ten thousand ETCs. It happens that one of these mitochondria does not have enough Complex IV, the last electron carrier in the ETC. As a result, in this particular mitochondrion, oxidative phosphorylation slows down, and electrons back up in the ETCs. This causes the electrons to escape and form superoxide free radicals. This mitochondrion is in jeopardy of damaging itself beyond repair. The logical solution to this scenario would be to manufacture more Complex IV. But how does the mitochondrion signal that it needs to build more Complex IV? The signal seems to be the free radicals themselves. Although the free radicals can inflict much damage, they can also control the action of "redox-sensitive" transcription factors, which are activated in response to oxidation by free radicals. In turn these transcription factors go on to alter gene activity to make more Complex IV.

Some readers might ask, "How does the cell know to interpret this free-radical signal as a message that more Complex IV is needed?" After all, low energy demand or a shortage of oxygen supply can also produce free radicals as discussed earlier, where neither situation would be improved by building new complexes. The cell does this by putting the free-radical message into context, just as we humans do in any conversation. In order to put a message into the context of what's happening, it needs other pieces of information. For example, in this case it might be the levels of ATP. If there were too few Complex IV entities in the mitochondrion, then ATP levels would fall (the ETC would get bunged-up as electron flow slows down). So the burst in free radical would be laid on a backdrop of low ATP, signaling the transcription factors to activate the genes for Complex IV production. On the other hand, if the cell detected high ATP levels, the burst in free radicals might signal that the proton gradient needed to be dissipated (and possibly more uncoupling proteins needed to be produced—more on this later).

Assume for a moment that all the genes are in the nucleus. The free-radical message arrives, and the nucleus sends orders to make more Complex IV. It then tags these new proteins with other proteins (an address tag) so that they can find their way back to the mitochondrion. However, the only thing these address tags do is tell the proteins that their destination is a mitochondrion, but they can't tell which specific mitochondrion is in need of the new Complex IV. This is analogous to me sending a package to a friend in a city when I don't know the specific address—there is a very slim chance the package would get to my friend. Further, considering that mitochondria are in a constant state of flux (being degraded, splitting into two through fission, or merging into one through fusion), the system wouldn't be too efficient even if the nucleus could tag the newly constructed proteins with a specific address; that address might no longer exist!

So all those new Complex IV entities would get distributed evenly to all one thousand mitochondria in the cell. The mitochondrion that needed it—the one that sent the original signal—doesn't get nearly enough Complex IV, while the rest receive too much (and send a corresponding message back to the nucleus instructing it to *stop* producing more Complex IV). It's easy to see the problems with this situation. Here's what's most important: When the mitochondria are not in control of their own destiny, the entire cell will inevitably have problems with energy production.

Now, let's consider the same scenario, but the genes for Complex IV are retained in the mitochondria (as they are in reality). When the free-radical signal to produce more Complex IV is sent, it goes straight to the mtDNA, which also happens to be in the immediate vicinity of the origin of the free-radical signal itself (so the response can also be very fast). The locally retained genes instruct the mitochondrion's own ribosomes to make more Complex IV, and these are immediately incorporated into the ETCs, clearing the backlog in electron flow and restoring efficient oxidative phosphorylation. Similarly, if and when the message to stop producing Complex IV is sent, it's specific to that mitochondrion and the response is immediate.

This rapid and localized response happens in each of the one thousand mitochondria in the cell, some requiring more Complex I, others more Complex III, while others need to dissipate their proton gradient. So while it is extremely costly for the cell to maintain tens of thousands of copies of mtDNA, the alternative (only a single copy in the nucleus) is ultimately far more costly and detrimental.

Allow me just one last detour down the deeply scientific path. The respiratory complexes are constructed from a number of separate protein subunits, and not all subunits are coded by the mtDNA. In fact, out of the forty-six subunits in Complex I, four subunits in Complex II, eleven in a Complex III dimer, and thirteen in Complex IV—a total of seventy-four different protein subunits—only thirteen are coded for by the mtDNA. The rest are coded for by the genes in the nucleus. The complexes of the ETC are a blend of proteins encoded by two separate genomes.

So, if this is the case, how does a mitochondrion, which only retained a fraction of the total number of genes needed for the ETC complex, maintain control over its own destiny? Well, it seems that the respiratory complexes assemble themselves around a few crucial subunits. Once these critical subunits have been embedded in the inner mitochondrial membrane, they act as a magnet that attracts the rest of the subunits to self-assemble appropriately. Thankfully, these crucial subunits are the ones encoded by the mitochondrial genes, and therefore the mitochondrion *is* able to retain control over the number of new complexes being built.

Because a cell has hundreds to thousands of mitochondria at any given time, the total number of these crucial subunits embedded in the inner mitochondrial membrane might remain fairly constant. So the nuclear genes and the overall rate of transcription could remain fairly constant, allowing individual mitochondria to control their own rate of oxidative phosphorylation while the nucleus could control the overall rate of energy production in the cell as a whole.

There is the fact, however, that *all* protein subunits in Complex II (which, by the way, is made of only four subunits) are coded for by the nuclear genes. However, this fact doesn't detract from the preceding discussion because both Complex I and Complex II pass on their electrons to Complex III. To a large extent, a mitochondrion can still control its own rate of oxidative phosphorylation by controlling the production of just Complexes I, III, and IV. Further, when you consider that Complex II is the only one that doesn't pump protons, it would make sense that at some point in the billions of years of evolution, the genes for all four protein subunits of Complex II were transferred to the nucleus, unloading a little bit of genetic burden and gaining a tiny bit of genetic efficiency.

Mitochondrial Mutations: The Beginning of the End

Mutations to mtDNA accumulate with the passage of time. Other than random mutations back to the original sequence, or recombination with an original DNA strand, the original sequence is lost, and this incorrect set of instructions will lead to defective proteins that don't function the way they are supposed to.

If the mutations affect any one of the many protein subunits in the mitochondrial ETC, the rate of free-radical leakage increases, and the situation can spiral out of control quite rapidly. Unfortunately, the odds are in favor of free radicals damaging the genes for ETC proteins. This is because mtDNA is stored next to the cell's primary site of free-radical generation. Mitochondrial DNA also doesn't have the protective histone proteins that nDNA does; its repair mechanisms are severely deficient; and there is no junk DNA (the genes are packed tightly together so that a mutation anywhere is likely to result in a negative effect). So it's only a matter of time before these genes are damaged, which would undermine the functioning of the ETC and oxidative phosphorylation.

A Radical Signal for Death

Intuitively, it seems that the rate of free-radical leakage from the respiratory chain would correspond with the rate of respiration—but it does not. Of course, energy demand and use, uncoupling, and other variables will be contributing factors to the rate of leakage (factors tied to the rate of respiration), but ultimately it depends on the availability of electrons (and oxygen).

We know the primary cause of mitochondrial damage is the free radicals generated by the mitochondria themselves. Current evidence suggests that the majority are generated by Complexes I and III (Complex I seems to generate free radicals if there is too much fuel relative to demand, and Complex III seems to generate them if ATP isn't being used up fast enough).

During normal oxidative phosphorylation, 0.4–4.0 percent of all the oxygen consumed is converted in mitochondria to superoxide free radicals. Superoxide is transformed to hydrogen peroxide (H_2O_2) by superoxide dismutase. H_2O_2 is then converted to water by glutathione peroxidase (one of the body's primary antioxidant enzymes) or peroxiredoxin III. However, when these enzymes cannot convert superoxide free radicals to H_2O fast enough (or when superoxide generation greatly increases for one reason or another), oxidative damage occurs and accumulates in the mitochondria.

In laboratory studies, superoxide has been shown to damage the iron-sulphur cluster that resides in the TCA entity called *aconitase*. This exposes iron, which reacts with H_2O_2 to produce hydroxyl radicals. Further, nitric oxide (NO) is produced within the mitochondria by mitochondrial NO synthase, and also freely diffuses into the mitochondria from the cytosol. NO reacts with oxygen to produce another free radical, called *peroxynitrite*. Together, these two radicals as well as others can do great damage to mitochondria and other cellular components.

However, all this depends on the availability of both fuel and oxygen. For example, consider a person in a developing country during a period of famine. This person has a shortage of fuel and, therefore, hardly any electrons flowing down the ETC. Even though there might be plenty of oxygen available, very few free radicals leak simply because of a lack of electrons.

Then consider a well-fed elite athlete in training. This person's muscle cells have plenty of fuel, but also a high demand for energy. Electrons flow smoothly down the ETC to oxygen, and relatively few free radicals leak because ATP is constantly used up.

However, what about a well-fed sedentary person? In this case the mitochondria have plenty of fuel, but the cells don't use the ATP that's been generated. ATP levels remain high with little turnover. With this low demand for ATP, the ETCs become backed up with excessive electrons. There is still plenty of oxygen as well as an abundance of highly reactive electrons, so there is a high rate of free-radical leakage. This burst of free radicals will exceed the built-in antioxidant defense system and oxidize the lipids in the mitochondrial membranes. This oxidation releases cytochrome c (which normally transfers electrons from Complex III to Complex IV) from the inner mitochondrial membrane and into the intermembrane space. At this point, electron flow down the ETC is completely stopped. The upstream sections of the ETC become full with electrons, and these electrons continue to leak and form more free radicals. Once this stress crosses a threshold, pores in the outer mitochondrial membrane open up and initiate the first steps of cellular suicide.

Dead or Alive: Mitochondrial Control over Life and Death

After energy production, perhaps the next most critical function of mitochondria is regulating death. When cells become worn out or damaged beyond repair, they are forced to commit cellular suicide, or apoptosis. If the mechanisms regulating apoptosis fail, the one serious consequence is cancer, which is why apoptosis is critical for the integrity and organization of multicellular organisms. This process is controlled by the mitochondria.

Genetically, multicellular organisms are composed of identical cells that perform dedicated tasks for the benefit of the organism as a whole. This concept is unique because all life has an innate desire to survive at all costs. So then, why and how did cells in a multicellular organism get to the point where they obediently follow directions— in most cases—to commit suicide for the greater good of the whole organism? The process likely evolved over hundreds of millions of years, and the enforcement of such altruistic behavior became the domain of the mitochondria, without which multicellular organisms would develop massive amounts of tumors everywhere and likely succumb to cancer very early in life.

In order to live in a community, and receive all the benefits that come with it, individuals need to commit to living in a manner that benefits the greater good, and some self-serving interests need to be forfeited. Unfortunately, cancers are common because selfish cells occasionally escape surveillance and skirt their death sentence. These cancer cells replicate feverishly without regard to the consequences of their actions on the community. In an obvious twist of fate, however, these delinquent cells eventually bring about the death of their own community, and in the end, as a result, themselves.

To prevent this, the mitochondria have evolved to play a central role in the civilized death of a cell. They do this by assimilating various signals from different sources. If the overall picture painted by the various signals indicates that the cell is no longer functioning properly, or within the confines of the greater good, then the mitochondria initiate the cell's suicide program. It begins with the activation of certain membrane receptors and pathways that involve another organelle called the *endoplasmic reticulum*, in addition to the mitochondria. A central event in many forms of apoptosis is activation of the mitochondrial apoptosis channel (mAC) by certain stimuli. Opening of the mAC causes the mitochondrial outer membrane to become highly permeable, and it therefore loses its electrical charge and proton gradient. This leads to a sudden burst in free radicals that oxidize various lipids of the inner membrane. For example, when cardiolipin becomes oxidized, it can bind to Complex IV, which is in turn released from its position in the inner membrane, shutting down the ETC.

This free-radical burst also releases cytochrome c (and other molecules) that joins with other components in the cytoplasm to form the "apoptosome," which then activates the enzymes of cellular death, the *caspases*. Remember, cytochrome c is responsible for shuttling electrons from Complex III to Complex IV. Under normal circumstances, it is secured to the outside of the inner mitochondrial membrane. However, once free in the cell, it binds to several other molecules to form a complex that activates the caspases.

The release of cytochrome c from its position on the inner membrane is seen as a critical step in apoptosis. Essentially, it pushes the process of apoptosis beyond the point of no return. What's interesting, however, is that simply injecting cytochrome c into a healthy cell will also cause it to die; this is a perfect example of the wise saying, "A little bit of knowledge is a dangerous thing." There are two components of the ETC that are not complexes themselves: CoQ10 and cytochrome c. CoQ10 is an amazing therapeutic natural health product, and supplementation has been shown to be beneficial for many health conditions. However, if we assumed the same for cytochrome c (thinking it would help shuttle electrons to Complex IV), we'd be killing ourselves (and this is the reason why there are no cytochrome c supplements).

Once the caspase enzymes are activated, they break down the cell in an orderly fashion. As the cell shrinks and then fragments, its organelles remain relatively intact and enclosed by membranes. Neighboring cells or macrophages safely digest the fragments and recycle the components for reuse. When executed properly, apoptosis is a well-orchestrated process of cellular self-destruction. Silently, in the human body, about ten billion cells are lost by this process every day.

The regulation of apoptosis is complex. Many checkpoints must be triggered before the apparatus of death is put into motion. Virtually all these steps are opposed by other proteins that counteract apoptosis, thereby preventing a false alarm. However, once the caspases have been activated, there is little chance of stopping the process. There is no doubt that there are thousands of ways a cell can be told to die. For example, activated immune cells send chemical signals to initiate apoptosis in cancer cells, DNA mutations from UV radiation, environmental toxins and pollutants, viruses and bacteria, various physical stresses and trauma, and inflammation (to name a few). However, all these diverse triggers activate the caspase cascade. In other words, all these signals somehow converge at the stage of the caspase enzymes; these enzymes, in turn, are activated by the burst in free radicals that follow the depolarization of the inner mitochondrial membrane and the release of cytochrome c.

Numerous studies show the value of apoptosis outside of controlling cancer growth and balancing cell division. It's also a key occurrence that happens all throughout nature. For example, during embryonic development in humans, vast amounts of neurons die in waves. In some areas of the brain, more than 80 percent of the nerve cells formed during the early phases of development disappear before birth (a rate similar to the loss of oocytes from embryonic development to birth). The death of all these neurons allows the brain to be "wired" with great precision. Functional connections are made between specific neurons, enabling the formation of neuronal networks, while others are eliminated. When some of these connections are not eliminated, there may be some unusual connections between different areas of the brain that normally do not communicate directly with one another. The result may explain some cases of autism, where some "higher functioning" individuals on the spectrum see colors and textures when reading numbers, or where specific numbers are connected to specific emotions.

An example from Lane's book describes how we don't form distinct extensions from our hands to form fingers. Instead, after the hand develops, apoptosis takes hold to eliminate the cells in between the digits. When this fails, the result is "webbed hands." The shaping of the body is brought about by subtraction, not addition.

By contrast, necrotic cell death, or *necrosis*, is where the cell swells and ruptures, organelles disintegrate, and inflammation tends to occur. This process can also begin with the opening of a channel in the inner mitochondrial membrane of the mitochondria called the *megachannel* (also called the *mitochondrial permeability transition pore*, or mPTP). And there is a more recently discovered third mechanism of death that falls between apoptosis and necrosis, aptly called *necroptosis* (a programmed form of necrosis), which can involve the mPTP.

The study of cell death is complex in itself, and it is even further complicated by studies showing that apoptosis is closely followed by necrosis, and vice versa, so it seems as if these separate cellular processes happen almost simultaneously. However, regardless of the pathway taken to cell death, in each case, it's the mitochondria that play the central role.

Discarded Theories of Aging

No matter the life span, all animals age in a similar fashion, just at different rates. Those with a fast metabolic rate age more quickly and yield to degenerative diseases in a short period of time. Those with a slow metabolic rate will follow the exact same chain of events, but just spread over a longer period of time.

In general, the larger the animal, the slower the metabolic rate, and the longer the life span. The exception to this is birds, as Lane discusses, because they have a fast metabolic rate but also a long life span with a lower risk of age-related degenerative diseases. Their mitochondria leak far fewer free radicals, and this fact—as I'll discuss shortly—has direct implications for aging, the risk of degenerative diseases, and death itself.

If we were to plot animals on a life-span graph based on their resting metabolic rate, on average, a bird will outlive a mammal with a similar resting metabolic rate by a factor of three or four (at least). In his book, Lane compares a pigeon to a rat. Both have similar resting metabolic rates, but the pigeon will live about thirty-five years while the rat will only live three or four years. However, a pigeon hasn't slowed down its pace of life to live that much longer than a rat—it maintains the same pace of living. Lane also points out that we humans live longer than we "should." We apparently live three or four times longer than other mammals with comparable resting metabolic rates.

It's also important to dissociate aging and degenerative diseases. While most warm-blooded mammals suffer an increasing number of degenerative diseases as they age, not all do, which is a small relief, for it helps us imagine a long life with the potential for little or no degenerative diseases. A long, healthy life—isn't that what we all want?

If and when a mammal does develop these diseases, the development is not dependent on a fixed passage of time, but instead is relative to the life span of the mammal—and that itself is somewhat fixed for each species. For this reason, rats are used as models for human disease in the laboratory—they suffer similar diseases of old age, but all within two or three years (not over decades). They can get diabetes, become obese, and develop cancers, heart disease, blindness, and dementia. Many birds also suffer from these diseases, but they only do so after a couple decades, which is why they aren't used as models for humans in laboratories.

Interestingly, long before I decided to write this book, I wrote an essay during my first year in medical school. The topic was a theory I had put together—without any evidence, I should add—that death and age-related degenerative diseases were not "normal" and humans are, in fact, meant to live indefinitely in great health. The premise was that death and disease were an invention or creation of evolution to ensure the survival of the human species at the expense of the individuals. The faster we age and die, the faster we must reproduce. If we didn't age or die, there would be no motivation to reproduce; lack of aging or death would put the survival of the species as a whole at risk—we would not be well adapted to a changing environment. Also, the longer a species lives, the longer the period of time between successive generations. The more "generations" a species can produce in a given period of time, the higher the likelihood of a beneficial combination of assembled genes (from mother and father), and the greater the likelihood that a random beneficial mutation is introduced to the gene pool—and this increased gene diversity improves the likelihood that some within the population will survive any given change in our environment. Hence, in my theory, aging took hold as a beneficial "mutation" that motivated us to reproduce every fifteen to thirty years. I know there are countless holes in this theory, which is probably why I was barely given a passing grade. Yet who would have guessed that a dozen years later I'd be writing this book with a related train of thought, but this time—learning from my mistakes—backed by significant scientific evidence spanning decades and many academic disciplines.

There have been countless theories on why we age. For example, the Endocrine Theory says that aging is caused by a fall in the circulating levels of a hormone, such as testosterone or estrogen. However, is this really a *cause* of aging, or just a consequence? *Why* do levels of these hormones start to fall in the first place?

Another is the Wear and Tear Theory, which stipulates that cells naturally just deteriorate over many years. Since this makes sense, it is a popular theory, but *why* do cells deteriorate? Why do different species accumulate wear and tear at such different rates?

What about the Telomere Theory? The telomeres—the "caps" at the end of chromosomes that slowly wear down over our lives—exhibit such varying patterns across species and even within different tissues of a human body that they can't possibly be the primary *cause* of aging. Besides, there are just too many holes in the theory at this point.

If you always ask "why," it'll be evident that most theories fail the test of causality or fall victim to assuming what they're attempting to prove. A good theory must overcome a number of apparent discrepancies, paradoxes, and gaps in logic. It must also explain any contradictory observations between different species. This brings us to the Free-Radical Theory, which says that the damage from free radicals is what causes aging and determines life span; this does indeed explain many discrepancies and paradoxes, even why birds live so much longer than other mammals despite their fast metabolic rate. While this theory recognizes that free radicals come from the mitochondrial ETCs, it also recognizes that those free radicals enter the body from outside sources.

As with all theories, the Free-Radical Theory has been challenged repeatedly, and with each challenge, the theory has changed shape and become stronger. However, this theory fails to explain the Exercise Paradox—the observation that athletes seem to live longer and healthier lives, although they consume far more oxygen (and generate far more free radicals) than sedentary individuals. This theory also fails in its application. If free-radical leakage from the ETC causes aging, then logically it makes sense that boosting the antioxidant defense system would prevent the damage in the first place and extend life span. Therefore, mammals with a long life span must inherently have a better antioxidant defense system. Based on this, birds must have high levels of antioxidants, and rats very little. It's therefore plausible that if we want to live longer and healthier ourselves, all we have to do is boost our own antioxidant protection. However, subsequent studies have disproved this theory. Birds have very low levels of antioxidants (and live a long time) and rats have an abundance of them (yet have very short lives). Further, antioxidant supplementation has failed to extend life span in laboratory settings. For several decades, researchers have administered countless antioxidants to all sorts of deteriorating biological systems without success. At best, they might reduce the risk of certain diseases, or even improve symptoms of other diseases, but none have ever lengthened maximum life span. Of course, an argument could be made that maybe the dose of the antioxidant was wrong, or the specific type was wrong, or the distribution was wrong, or the timing was wrong. The fact is, however, that antioxidants just aren't the cure-all we previously thought they might be. A number of independent studies have advocated that there is actually a negative correlation between endogenous antioxidant levels and maximum life span. Simply put, the higher the antioxidant concentration, the shorter the life span (and, in fact, numerous studies have shown oxidative stress can actually *prolong* life span).

Thankfully, the dietary supplement industry has recognized this and we no longer hear much about antioxidants the way we used to. Even up to a few years ago, ORAC (oxygen radical absorption capacity) —a measure of a substance's antioxidant power in a test tube—was flaunted by savvy marketers as the cure for everything. But we know that we can't measure the effect of substances in a biological system by using test tubes as the measuring device. Newer research is showing that many "antioxidants" still do have therapeutic benefits; however, the mechanism is not due to their antioxidant effects, but due to their ability to modify the expression of certain genes related to the conditions being targeted (i.e., turning relevant genes on or off). While I find this extremely fascinating (and I will discuss this in passing), a full discussion is beyond the scope of this book.

However, it remains that it's very difficult to target antioxidants to the mitochondria (and this is an active area of pharmaceutical research). This has given birth to the Mitochondrial Theory of Aging, which itself has transformed significantly since its inception in the early 1970s. It is currently the strongest theory that explains aging;

it also explains why we suffer from degenerative diseases later in life, explains the Exercise Paradox, and sidesteps all the other pitfalls that have consumed the other theories.

The Mitochondrial Theory of Aging

A modern version of the Mitochondrial Theory of Aging was initially put forth by Anthony Linnane, an Australian professor and scientist, back in the late 1980s. The theory has gone through some further modifications since then, but it essentially ignores exogenous sources of free radicals. The theory's main point is that the mitochondria are the body's main source of free radicals that are linked to aging.

Free radicals don't damage the cell as much as we think. We produce a number of antioxidant enzymes that mop them up, and of course, if the cell does get damaged, there are repair mechanisms, and these are constantly at work. But the free radicals linked to aging do specifically damage the mitochondria, and especially their vulnerable DNA, which doesn't have the repair mechanisms the rest of the cell has. When the damage accumulates faster than a mitochondrion can repair itself, it becomes dysfunctional, the first step in aging. In essence, this theory says the mitochondria are the "biological clock." The chain of events would look like this: Free radicals escape the respiratory chains (in different scenarios, some already discussed) and attack the mtDNA in its immediate proximity, which leads to mutations that might compromise mitochondrial function. As the mitochondria start to weaken and eventually die, the functioning and viability of the cell as a whole decline. As cells lose their ability to produce energy, they commit apoptosis, which then compromises the functioning and viability of the tissue or organ.

As random mutations in mitochondria accumulate, a *bioenergetic mosaic* develops—where cells all produce a drastically differing amount of energy, depending upon their degree of mitochondrial damage (some produce relatively little energy, some produce moderate amounts, and some produce a large amount of energy). In a healthy toddler, we do not see this mosaic because energy production is high in nearly all cells. However a noticeable *mosaic effect* develops after about age forty, and the extent of this effect depends upon the rate of bioenergetic aging in a particular tissue.

From a bioenergetic perspective, some tissues appear to age very quickly, others at a moderate rate, and some quite slowly. This effect also illustrates why the biological age of two people could vary significantly, even though they might be the same chronological age.

Linnane's theory holds that this mutation-driven bioenergetic decline is a major factor in the degenerative diseases and general frailty of old age. Recent research from numerous disciplines has converged on the mitochondria as the center of cellular aging, giving more weight to the theory, and enhancing it. If the Mitochondrial Theory of Aging is correct, the foundation of cellular vitality lies in the mitochondria.

However, through evolution, the rate of free-radical leakage has been set at the optimal level for each species. A bird, as Lane discusses, leaks fewer free radicals and, therefore, has a long life span despite a fast metabolic rate. So then he asks the question, why don't all species have well-sealed mitochondria? A rat would surely benefit from reducing free-radical leakage in the first place, rather than spending copious amounts of resources to produce large quantities of antioxidants. Makes sense to me. However, the answer is what makes the Mitochondrial Theory of Aging radically (pun intended, as you'll see) different from the Free-Radical Theory.

Remember the reason why we have a separate copy of DNA in the mitochondria? It is to balance the requirements of oxidative phosphorylation, because an imbalance in the components of the ETC can lead to unproductive respiration and free-radical leakage. By keeping a local set of important genes, each mitochondrion retains control over its own respiration, based on its own need (and not that of other mitochondria).

Also remember that the signal to produce more componentry for the ETC comes from the free radicals themselves. This could be why a rat might need the excessive free-radical leakage: If it were to have more tightly sealed mitochondria, its free-radical signal would be weakened by the high amounts of antioxidants and so require a more refined detection system.

So just as we typically view a forest fire as a bad thing (just like free radicals), in moderate amounts, it serves a vital role in the maintenance of the ecosystem (fires break down organic material in a matter of minutes to hours, which would otherwise take years or decades to decompose; they clear the ground to make room for new growth; and some plants, such as the jack pine, require fire to melt the protective resin that prevents them from germinating). At the same time, firefighters (the antioxidants) do have their role, but if they were too vigilant, a forest would never get the chance to regenerate and renew itself. Then there is fighting fire with fire (such as using a prooxidant, or promoting oxidation). When used properly, it can be a valuable tool (i.e., the controlled use of oxidants can be used therapeutically, such as in treatment of cancers, high-dose intravenous vitamin C treatment, and so on). However, used improperly, it can become an uncontrolled fire itself.

While it's not known exactly how free-radical signals work, we do know that the system behaves like a thermostat, and it requires a certain degree of fluctuation in free radicals. If the rate of free-radical leakage from ETC didn't fluctuate, there would be no self-correction (just like if temperature in a room didn't fluctuate, self-correction wouldn't happen).

However, when the free-radical signal fails, or indicates an uncorrectable problem, then the free radicals initiate the apoptosis program. When this happens in just one or a few mitochondria, the signal isn't strong enough for the cell to commit apoptosis. However, if a large number of mitochondria collapse simultaneously, then the signal crosses the threshold and the cell knows its time has come. So whereas the Free-Radical Theory and the early renditions of the Mitochondrial Theory of Aging both suggested that free radicals were purely detrimental, the modern version of the Mitochondrial Theory of Aging appreciates that free radicals carry out an essential signaling role.

Nonetheless, this doesn't detract from the point that free-radical leakage *is* correlated with aging and life span. It's known that mitochondrial mutations in the "control" region of mtDNA accumulate with age. This accumulation is significant to note because mutations in the control area in one cell can often spread to all the cells in the tissue. When a mutation is in this control region, it might affect the binding of transcription or replication factors—but it doesn't affect gene sequence. Depending on the resulting effects of this mutation, it will tend to copy itself either more often or less often. So, for example, if the mutation makes a mitochondrion respond more sluggishly to a given signal to replicate, when the signal to divide comes, the "normal" mitochondrion would divide and replicate, but the defective mitochondrion might not or might to a far lesser degree. Relative to the normal mitochondria, the number of the defective mitochondria would continually decrease, and eventually they would be displaced completely with the regular turnover of cellular constituents. If the mutation made the mitochondria swifter in their response to the same signal, their DNA would proliferate, and eventually they would displace the "normal" mitochondria in the cell. It's important to note that such mutations are more likely to take over all cells in a tissue if they're not particularly detrimental to mitochondrial function (e.g., the ETC components are still normal); otherwise the cells would just die.

By contrast, mitochondrial mutations in the coding regions can be amplified within particular cells, but rarely spread beyond 1 percent of the cells in a tissue. The reason is that these mutations will likely affect mitochondrial respiration because they code for the critical protein subunits of the ETC. The result would be an increase in free-radical leakage, but unlike any normal circumstance, the signal to build more new complexes would not correct the deficit because all the new proteins would be defective as well. This sounds catastrophic, doesn't it? Thankfully, under the modern version of the Mitochondrial Theory of Aging, it is not. Instead, the defective mitochondria signal their deficiency to the nucleus, which then allows the cell to adapt.

This signaling from mitochondria to nucleus is called the *retrograde response* because it's opposite to the normal chain of command (from the nucleus to the rest of the cell). The overall intention is to correct the metabolic deficiency. Retrograde signaling switches energy generation toward anaerobic respiration (energy production without using mitochondria and oxygen), and this switch stimulates the genesis of more mitochondria, or mitochondrial biogenesis, which also protects the cell against future metabolic stress, which, in the long run, is the only option a cell really has to correct any bioenergetic deficiency.

The population of mitochondria is in continuous flux. Mitochondria will divide if the energy deficit is fairly mild, and cells will amplify the least-damaged mitochondria—simply because these work the best—and defective ones will eventually die out. The mitochondria that die will be broken down (in an orderly mitochondrial version of apoptosis, called *mitophagy*), and their components will be recycled. Ultimately, the most damaged and dysfunctional mitochondria are continuously eradicated from the population. In this fashion, most cells can theoretically extend their life almost indefinitely by persistently correcting the deficit.

You might think I've just revealed the Fountain of Youth. Perhaps, but if it were as simple as that, my physicality would still be that of a physically fit twenty-something-year-old (at least by choice), and it's definitely not. While the most destructive mitochondrial mutations can be eliminated, there is no way of restoring them to their original youthful state—at least not naturally. Maybe in the future we'll be able to extract undamaged dormant mitochondria from our stem cells and have them infiltrate every other cell, but I'd say we're a few years away from that. What does seem possible, however, is the ability to slow the aging process, and delay or even prevent diseases connected to age-related mitochondrial decay.

As we age, the cells rely more and more on defective mitochondria. Things don't spiral quickly out of control, however. The cell and its mitochondria retain control by adapting their behavior, establishing a new equilibrium. Most studies searching for evidence of damaged proteins, lipids, and carbohydrates have failed to find any serious difference between young and old cells. Instead, we find evidence that the spectrum of operative genes is what's affected, and this hinges on the activity of transcription factors. The activity of some of the most important transcription factors depends on their redox state (whether they're oxidized or reduced), many of which are oxidized by free radicals, and reduced again by dedicated enzymes. It's this delicate balance between the two redox states that determines their activity. (The field of *redox medicine* is also a growing discipline of study.)

Redox-sensitive transcription factors can behave like radar, alerting the cell to imminent threats and enabling it to take appropriate action. Their oxidation initiates the changes that will prevent any further oxidation. For example, Nrf-1 is a nuclear (meaning "from the nucleus") transcription factor that coordinates the expression of genes needed for mitochondrial biogenesis. If the conditions in the cell become more oxidized, Nrf-1 is activated, and in turn it stimulates mitochondria to divide, in an attempt to restore balance in the redox state. Nrf-1 also induces the expression of numerous other genes that protect the cell until more mitochondria are generated.

The more oxidizing the cell's internal environment becomes, the more these redox-sensitive transcription factors shift the activity of nuclear genes away from the day-to-day administrative work and toward crisis management that protects the cell from stress. This shift establishes a new equilibrium in the cell, one where more resources are dedicated to crisis management instead of their original tasks. This new equilibrium can be stable for years, even decades. Only the least-damaged mitochondria tend to replicate, so there is usually no obvious sign of mitochondrial mutations or damage. We do not die right away, but we might notice we fatigue more easily, or take longer to recover from illnesses, or forget more things.

The Mitochondrial Theory of Aging explains why we don't see the spiraling, catastrophic damage predicted by the Free-Radical Theory or the other original mitochondrial theories. Free radicals are used to signal danger, which allows the cell to adapt. The Mitochondrial Theory of Aging also explains why the cell doesn't have more antioxidants than it needs; if there were too many, it wouldn't be sensitive to changes in the redox state. Without free radicals, the whole system would fail and the mitochondria would fail to adapt to their changing environment or demands. Surely this would lead to a high rate of mutations and result in a quick end to us all.

Unfortunately, after numerous decades of constantly adapting to a new equilibrium, cells eventually run out of healthy and normal mitochondria. After this happens, when the cell signals mitochondria to replicate, there is no choice but to amplify the defective mitochondria. Ultimately, cells are overtaken by these defective mitochondria.

Interestingly, if we examined a dysfunctional organ or tissue, we wouldn't actually see an abundance of cells with defective mitochondria—we would only see a few affected cells at any given time. When cells finally get to the point where they are packed full of defective mitochondria, they are given the signal to remove themselves from the community by apoptosis. While this fact means we don't find high levels of defective mitochondria in aging tissues, it does result in the slow but steady loss of tissue density and function (osteoporosis or sarcopenia, for example)—the prerequisite for aging, disease, and ultimately death.

To make matters even more complicated, and to give a shred of hope, it's been recently shown that mitochondria are better at repairing damage to mtDNA than previously thought. The five to ten copies of mtDNA usually present in every mitochondrion means that, at any given point in time, a good copy of a particular gene is still in working order. That copy is used as a template for recombination (repairing) of the damaged gene. However, the significance of this discovery to the newest rendition of the Mitochondrial Theory of Aging is yet to be determined.

Nevertheless, beyond being consistent with the emerging scientific data, the modern Mitochondrial Theory of Aging provides profound insight into the pathology of age-related degenerative diseases, and how to prevent or maybe even cure them.

Extending Maximum Life Span in Mammals

Each species of mammal has a theoretical maximum life-span potential (MLP). While significant medical and public-health advances have produced staggering increases in *average* life span, the MLP of about 120 years for humans has yet to be extended. Nor has the species-specific MLP been convincingly extended in other mammals, the exception being studies on caloric restriction. The increase in MLP caused by caloric restriction, and the fact that calorie-restricted animals are biologically younger at any given age, suggests that at least some basic aging processes are decelerated by caloric restriction.

Thus far, only caloric restriction has been shown to extend MLP in nearly all species tested, including invertebrates, fish, and warm-blooded vertebrates such as mammals. The studies on caloric restriction have added even more weight to the Mitochondrial Theory of Aging. Why is this the case? Think back to the example I used earlier: a person in a developing country during a period of famine, who experiences a shortage of fuel, and therefore has hardly any electrons flowing down the ETC. Even though there might be plenty of oxygen available, very few free radicals leak simply because of the lack of electrons. However, in famine, there is also malnutrition. Caloric restriction differs from famine because a person might significantly lower the calories they consume, but they ensure that the foods they do consume are nutrient-dense. The result is that very few free radicals leak due to a reduced amount of electrons.

This concept also helps explain the opposite. Excessive caloric intake introduces an excessive amount of fuel into the body, and ultimately excessive electrons into the mitochondrial ETCs. An overabundance of electrons causes leakage at a very high rate, which might be why obesity (when a person consumes far more calories than they expend) is linked to countless degenerative diseases.

Although an increase in the MLP hasn't yet been achieved for humans, based on what I just discussed, there's a great deal of hope that it will happen soon. If indeed mitochondria are at the crux of aging, the fact that the best mitochondria with the best mtDNA are used as templates for the generation of more mitochondria, that the cell is better at repairing damage to mtDNA than previously thought, and that defective mitochondria are constantly eliminated, at least theoretically, the cell should be able to go on doing this indefinitely.

Degenerative Diseases and the Eventual End

We still don't know exactly what the signal for apoptosis is, but two related factors are probably involved: the percentage of dysfunctional mitochondria and the total ATP level in the cell relative to its demand. As a cell gets the signal for apoptosis, what happens to the tissue, and eventually the whole organ, is dependent on the type of cell it's composed of. If it's a type of cell that is regularly replaced by stem cells that have preserved its mitochondria in an untarnished state, there won't be any negative effect. However, if it's a type of cell that's typically irreplaceable, such as a nerve cell, then with each cell death, the tissue starts to atrophy and the remaining cells are under greater pressure to meet the functional demands of the organ. As the surviving cells get pushed closer to their own metabolic threshold, they are more likely to be negatively affected by the countless number of external factors that could place additional strain on them. As we age, the process accelerates as fewer and fewer cells are left to do the work of many. Remember, this explains why we don't see mitochondrial mutations spiraling out of control—defective mitochondria, and the cells that contain them, are constantly eliminated. However, the number of functional cells in any particular organ will decrease, which is known as atrophy.

This is how degenerative diseases take hold. As the quality of beta cells in the pancreas decreases, insulin levels fall off; as the heart loses its muscle cells, its contraction becomes less efficient; as the neurons in the brain start to die off, dementia sets in. In each case, there is a threshold. Losing a few cells in the heart is unlikely to result in heart failure, but lose enough and the function of the heart will be compromised.

If you think this discussion of the degenerative disease process sounds very familiar to the discussion of aging, you're right. The process is the same and shows how aging and degenerative diseases are linked. More importantly, if we could target the underlying process of aging, we could theoretically extend our MLP *and* delay all age-related degenerative diseases.

What's important to note is that while the rate of free-radical leakage correlates very closely with life span, it's how free-radical production influences the threshold for apoptosis that really matters. Some species, such as rats, have cells that leak large amounts of free radicals quickly. These cells are closer to their threshold, so it won't take long before they get the signal for apoptosis. For humans, it will take many, many more years for our cells to reach their threshold. If we could just slow down the rate of free-radical leakage from the mitochondria even further, we could significantly delay the onset of degenerative diseases or perhaps even eliminate them altogether. At this point in time, improving mitochondrial function and slowing their decay appear to be the most promising—and realistic—way to address both degenerative diseases and aging. It's spine-tingling to think that we're so close to potentially finding the answer to a long and healthy life.

The pharmaceutical industry spends many billions of dollars annually for research, but nothing more than symptom management comes of it. The paradigm under which that industry operates is likely one of its many problems. Drugs are almost exclusively used after a disease has manifested its physical symptoms; they are rarely, if ever, used to prevent diseases in the first place. If it's true that the quality of our mitochondria is the single most important factor in aging and degenerative diseases—and if we can't turn back the clock on our mitochondria—prevention should start in childhood.

As I noted previously (see "Discarded Theories of Aging" on page 43), even the dietary supplements industry is on the wrong path with all its marketing of antioxidants. The antioxidant craze promotes these supplements as the cure for most ailments, and although it seems to be losing some steam, *antioxidant* is still a buzzword thrown around excessively to hopeful consumers. Also as mentioned earlier, while antioxidants do have some benefits in certain diseases according to *some* studies, other studies have found that large amounts can potentially do some harm. Just because they're marketed as natural and healthy, this doesn't mean it's good for you to use them indiscriminately or in excessive amounts. If you mess with the mitochondrial thermostat, the cell can't calibrate its response to stress appropriately. In the long run this can't be good, and it undermines Nature's protective processes. This mitochondrial thermostat also explains why while antioxidants might extend life in a sick population (relative to people with the same condition who are not receiving antioxidants), they fail to extend the MLP of a species. Antioxidants are likely beneficial to extracellular components, at membrane surfaces, and maybe even in the cytoplasm of our cells, but it is highly unlikely they will be able to quench the free radicals leaking into the mitochondrial matrix.

Yet all the expanding knowledge about mitochondria gives us new hope and insight for treating illnesses. If all the genetic and environmental factors that lead to age-related degenerative diseases converge at the mitochondria, we just need to focus on one organelle.

While newer research is revealing the intricate interaction between mitochondria and other organelles, such as peroxisomes and endoplasmic reticulum, we seem to be one big step closer to targeting the underlying mechanism behind many diseases at once.

It's Getting Hot in Here: Uncoupling the Proton Gradient

Lastly, no discussion on mitochondria would be complete without understanding their role in heat production. The proton gradient is not just used to produce energy; sometimes it is uncoupled from energy production and the gradient is dissipated as heat. Specifically, the electron flow and proton pumping continue normally, but the protons don't flow back through the ATPase, and thus ATP is not produced. Instead, the protons pass back through other pores in the membrane (descriptively called *uncoupling proteins*, or UCP), where the energy contained in the proton gradient is released as heat.

In fact, this process is how warm-bloodedness evolved, and it is the source of nonshivering thermogenesis (production of heat without shivering), which happens predominately in what's called *brown adipose tissue* (or more commonly known as *brown fat*). In contrast to cold-blooded animals such as reptiles, warm-blooded birds and mammals can generate their heat internally, which is called *endothermy*. In fact, this is the definition of "warm-blooded"—the ability to generate heat internally (the actual temperature of blood can be similar in warm- and cold-blooded animals).

Many organisms (including snakes, sharks, and even some insects) are endothermic. They generally use their muscles to generate heat during physical activity. In mammals, such as humans, muscles contribute to endothermy by shivering in intense cold or during vigorous physical activity. However, birds and mammals can also generate heat based on the activity of their internal organs, such as the brain and heart.

How this came to be is another interesting discussion you can read in Lane's book, but the conclusion is that endothermy, outside of the obvious advantages of physical performance (i.e., warm muscles react more quickly) and adaptability to cold environments, protects mitochondria from damage by maintaining electron flow during times of low energy demand.

But how? If ATP isn't used due to low energy demand, ADP becomes scarce and the ATPase grinds to a halt. This is when electrons running down the ETC are inclined to escape and react with oxygen, producing the destructive superoxide free radicals. Let's go back to our analogy of a hydroelectric dam on a river. During times of energy low demand, water flow (protons) through the turbines (ATPase) is reduced and the reservoir behind the dam is at risk of flooding (free-radical generation). The risk of flooding can be minimized by opening the overflow channels (UCP).

To illustrate further, let's look at an example of an athlete who just trained, gobbled down a meal, and then sat down to rest. The act of training required a significant spend in energy, but the reserves quickly get replenished as glycogen and fat are extracted from the food. There is little need to expend more energy, and the mitochondria fill up with electrons extracted from the food. From our previous discussion, we know this is a treacherous situation. When the ETCs become full with electrons (due to slow electron flow from lack of energy demand), the electrons can easily escape to form reactive free radicals, which can go on to damage the cell.

To minimize this situation, the athlete could get up and start moving again to use up some of that excess energy, but the other option is to dissipate that energy, which can keep the whole system from overflowing. This is accomplished through the use of UCPs, which act as the overflow valves or channels. This uncouples the proton gradient, so electron flow is no longer linked to ATP production. When protons travel through these UCPs, the energy stored in the proton gradient is dissipated as heat. By dissipating the proton gradient in this way, electron flow down the ETC is maintained because proton pumping can continue without overflowing the gradient. The result is less free-radical formation.

In resting mammals, up to 25 percent of the proton gradient is dissipated as heat. In fact, small mammals such as rats, and even human infants, need to supplement their normal heat production with brown fat. Brown fat has lots of mitochondria and lots of UCPs, and because nearly all the protons leak back through UCPs to generate heat, it becomes increasingly essential as the surface area to volume ratio of a mammal increases (smaller mammals and human infants lose heat much faster than larger mammals).

Being able to manipulate brown fat, UCP, and metabolic rate in general—while simultaneously minimizing free-radical production—has significant importance for preventing numerous health conditions (as I'll discuss in chapter 3, "Massage and Hydrotherapy," page 177). At the very least, manipulating brown fat would be a great way to help people prevent obesity. Nonetheless, I still find it captivating to think that without mitochondria and its uncoupling of the proton gradient, warm-bloodedness would have never evolved, and we'd likely all be enjoying a reptilian lifestyle, with all its limitations.

It's also interesting to note that this is how animals such as polar bears, who live in the extreme opposite of environments as camels (discussed on page 19, "A Game of Hot Potato: The Electron Transport Chain [ETC]"), survive. Polar bears live only in the northern Arctic, where they spend most of their time on ice floes. They are found in the United States (Alaska), Canada, Russia, Greenland, and Norway—areas that are viciously cold, with temperatures as low as –55°C and wind speeds of up to 50 kilometers per hour.

By having a thick layer of fat (also known as *blubber*) under their skin, the bears are not only great swimmers (because they use their fat and two layers of dense, oily, water-repellent fur to help keep them afloat), but with plenty of brown fat, they're able to generate lots of heat to keep themselves warm.

Because they have such large amounts of brown fat, about half of the food polar bears eat is used for the sole purpose of keeping them warm. The colder the Arctic gets, the more they must eat to keep warm. All this accumulated fat, fat consumption (mainly seal blubber), and fat-burning to keep warm means polar bears rarely need to drink water—instead they meet their water demand from their food and through burning the accumulated fat, which ultimately results in water production (at Complex IV, part of the ETC where water is generated, which is similar to camels). Research has shown that if you see a polar bear drinking water, it means it is suffering from extreme exhaustion and starvation. It's also interesting to note that with global warming and climate change and the resulting loss of their sea ice habitat, polar bears are having to spend more energy swimming farther distances offshore between ice floes to hunt seals, leaving less fat for heat production. Further, scientists have also observed that they are—in the absence of sea ice—foraging on land, where they have difficulty finding prey, and therefore they are eating bird eggs that don't provide nearly enough fat for them to sustain themselves.

Brown fat, according to Lane, also helps explain some differences in health risks between different racial groups. As I discussed, mitochondria occupy the crux of degenerative diseases and aging, and the degree of uncoupling has evolved differently between ethnic groups. When we look at an Inuit population in the Far North, for example, we see they have relatively large amounts of brown fat. It doesn't take a rocket scientist to see why—the constant exposure to cold temperatures necessitates that these individuals are able to produce vast amounts of heat to stay warm, just like the polar bears. Due to their large amount of brown fat, the Inuit mitochondria don't leak as many free radicals, and consequently, this population is known to have a low incidence of degenerative diseases, such as heart failure, which is so common in Western populations.

On the other hand, those of African descent, whose mitochondria have evolved in the blistering heat of the equatorial sun, would not benefit from excessive heat production, and therefore, they have relatively small amounts of brown fat. Their mitochondria are "tight" and more of the proton gradient is used to generate ATP and energy, not heat. Unfortunately, there is also a larger amount of free-radical generation—and studies show that African Americans have a much higher risk of degenerative diseases than most other populations. Exercise and physical activity are critical for individuals whose maternal lineage can be traced back to equatorial cultures (remember, mitochondria are inherited through the maternal line)—these individuals *must* ensure they are using up their ATP constantly. Of course, mitochondria are just part of the picture, as there are many other physiological, epigenetic, and socioeconomic factors that lead to higher risk of degenerative diseases in this population—but it is eye opening to see how at least some of these observations can be explained by looking at the differences in mitochondrial genetics.

I can't speak for anyone else but myself, but I find this stuff incredibly fascinating! I hope you do, too. So, now that you've been introduced to the history, evolution, and significance of mitochondria, let's discuss what we've discovered in terms of its involvement in human diseases.

CHAPTER TWO

The Dark Side of the Force

Health Conditions Linked to Mitochondrial Dysfunction

I n the previous pages we examined how energy production by the mitochondria is a foundation for health and well-being, and is necessary for physical strength, stamina, and even consciousness. We know subtle deficits in mitochondrial function can cause weakness, fatigue, and cognitive difficulties, and we know that certain chemicals that interfere with mitochondrial function are known to be potent poisons. Indeed, compromised mitochondrial function is now seen as one of the leading causes of a wide range of seemingly unconnected degenerative diseases, and even the aging process itself.

In this section, I will elaborate more on this dark side—the health consequences associated with defective or compromised mitochondria. I should make it clear, however, that my discussion will not be anywhere near comprehensive. In fact, only a fraction of what I've researched actually made it into this book; this is only the tip of the iceberg. The conditions I discuss are wide-ranging, which will hopefully give you an appreciation for the importance and central role that mitochondria play in our lives—in both health and sickness. Some of the diseases I cover here are genetic (and comprise the group of conditions collectively known as *mitochondrial diseases*), while others are acquired through epigenetics (from viral infections, exposure to environmental toxins, excess caloric intake, the natural aging process, and everything in between).

A Review of Bioenergetics

Bioenergetics is the study of energy in the human body, and it is critically important that we understand this concept, because problems with energy production and use are at the heart of many diseases where mitochondria play a central role (as I'll soon discuss).

While scientists know a great deal about ATP, there is a general lack of understanding by physicians and other health care practitioners about how to apply bioenergetics in practice. Let's look at just the heart for starters, which contains approximately 0.7 gram of ATP, just enough to fuel the heart to contract at a rate of about sixty times a minute, or about one contraction per second. At this rate, which is relatively slow for even "healthy" people, a heart will beat 86,400 times a day. So the heart needs 6,000 grams of ATP in a day—replenishing its energy pool ten thousand times! But how does the heart manage this incredible production of ATP?

We need to understand ATP first, before we can understand the process of its recycling. ATP is composed of three major chemical groups: *adenine* (a purine base), *D-ribose* (a pentose, or five-carbon sugar), and three phosphate groups. Energy is released to the cell when an enzyme removes a phosphate from ATP and converts the chemical energy stored in the bond into mechanical energy. After the phosphate is removed, what remains is ADP. You might recall from our earlier discussion in chapter 1 (see "ATP Synthase: Coupling the ETC with Oxidative Phosphorylation," page 26) that through the use of the ATPase in the inner mitochondrial membrane, the phosphate is reattached to ADP to re-form ATP.

As long as each cell is provided with two basic ingredients— electrons from food and oxygen from the air we breathe—this cycle occurs unimpeded millions of times per second in every cell of the body. This continual "recycling" keeps the cell fully charged with energy. However, if either of these two basic ingredients is in short supply relative to demand, cell function is compromised.

One good example is the oxygen deprivation experienced after a heart attack, or what the medical community calls a *myocardial infarction* (MI). An MI occurs when an artery that supplies fuel and oxygen to the heart muscle cells becomes blocked. The heart muscle continues to use energy at its regular pace, but it has no oxygen—a dangerous mismatch in supply versus demand.

While it's not possible to determine accurately if ATP pools are compartmentalized or if they flow around freely, scientific evidence strongly suggests that there are localized areas of higher concentrations to perform specific tasks (such as contraction in heart muscles or movement of ions across membranes). However, no matter where ATP is found, once the ATP releases its energy and is converted to ADP, it must be recycled back to ATP, where it once again leaves the mitochondria to where it's needed.

A small amount of ADP remains in the *cytosol* (the fluid part of the cell), where it is converted to ATP (instead of entering the mitochondria for recycling). This ATP is generally associated with cellular membranes and provides the energy needed to control the movement of ions across membranes.

If mitochondria produce 90 percent of the energy that the cell needs, how does the ATP get transported to the rest of the cell itself? ATP formed within the mitochondria must be moved back to the cytosol of the cell so that its energy can be used. At the same time, ADP from the cytosol must be moved back into the mitochondria to be recycled back to ATP. However, the mitochondrial membrane is impermeable to both ATP and ADP, so these compounds are "traded" across the mitochondrial membrane through the use of another enzyme called *ATP-ADP translocase*. This enzyme keeps ATP flowing out of the mitochondria to where its energy can be put to good use, and keeps ADP flowing into the mitochondria. This process is like any manufacturing plant that uses recycled materials —recycled paper, for example, where new paper is made from old paper. In order to keep the whole system running smoothly, the new paper has to be shipped out, so that it can be used and then make its way back to once again create new paper. If there isn't used paper coming back to be recycled, we run out of our raw material. If we don't ship out the newly created paper, it can't be used to create more of our raw material. We could use virgin paper pulp if we didn't have enough recycled material, but creating it is a long and intensive process that requires years to grow a tree and extensive resources to turn the tree into our raw material for creating paper. In a similar way, we need ATP to be used and broken down to ADP—we need this ADP to come back to the mitochondria and serve as our raw material to create new ATP. It's true we can build more ADP from scratch, but just like growing a new tree, it's a slow and inefficient process.

Food and Oxygen:
The Ingredients for Producing Energy

The basic requirements for life—the food we eat and the air we breathe—ultimately serve to provide mitochondria with the ingredients they need to produce energy. The most readily available and main source of fuel is *glucose*, which is a simple six-carbon sugar extracted from the food we eat. When we've had enough food to satisfy our immediate need for energy, the rest is then stored as *glycogen*. Most people don't eat constantly throughout the day, so this stored glycogen is drawn upon and broken down into glucose as needed. The first step in this process is called *glycolysis*, which takes place in the cytosol portion of the cell. Because much of this process occurs near the cell membrane, scientists believe this pathway is used primarily to generate ATP for the movement of ions across the cell membrane. And while glycolysis is capable of producing large amounts of ATP quickly, it cannot supply nearly enough energy to keep the cell functioning for long periods of time. Only two molecules of ATP are produced if we start the process directly from readily available glucose. If we start from the stored glycogen, three molecules of ATP are created.

During normal carbohydrate metabolism, the six-carbon glucose is converted to two three-carbon molecules called *pyruvate*. Pyruvate can then enter the mitochondria and participate in the second pathway of energy production: the TCA, or Krebs, cycle.

As long as the cell is supplied with sufficient oxygen, pyruvate is converted and broken down further by the TCA cycle, where the resulting compounds can enter the third pathway: the ETC (electron transport chain).

However, if the cell is deprived of oxygen, such as during overly strenuous physical activity or if there is significant blockage of an artery, then the TCA cycle doesn't function efficiently and pyruvate is converted to *lactic acid* (also called *lactate*). Lactic acid causes the pH of the cell to drop (meaning increasing acidity), which in turn signals to the cell that more energy is needed. However, if lactic acid levels build up too high, this causes cellular stress; on the macro level, we feel this as burning and pain in the case of physical activity, or chest pain (*angina*) that can occur if there is decreased blood flow to the heart (called *cardiac ischemia*).

So while glycolysis is essential and glucose is readily available for most of the general population, it's not actually the most efficient pathway of energy production, and glucose is not the ideal source of fuel—fatty acids are. Fatty acids are metabolized in a process called *beta-oxidation*, and the burning of fatty acids is responsible for 60–70 percent of all of the energy our cells create. This is where *L-carnitine* (discussed in more detail in chapter 3, "L-Carnitine," page 153) enters the picture. The inner mitochondrial membrane is impermeable to long-chain fatty acids, but the fatty acids must enter the mitochondrial matrix where beta-oxidation takes place. Interestingly, L-carnitine is the only molecule that can transport long-chain fatty acids into the matrix, and without L-carnitine, the body's ability to use long-chain fatty acids to create energy would not exist.

The product of beta-oxidation enters the TCA cycle (just like pyruvate does from glucose metabolism). The role of the TCA cycle is to remove electrons from fatty acids and pyruvate and package them into other electron-carrying molecules, such as NADH and $FADH_2$, which then enter the ETC.

The net result is that each molecule of glucose forms a total of thirty-eight ATP molecules (two from glycolysis, thirty-six from

TCA/ETC), but each molecule of a sixteen-carbon fatty acid called *palmitate* produces 129 ATP molecules. You can clearly see why healthy, well-functioning cells prefer fatty acids as their source of fuel.

Just from these numbers it's clear that without L-carnitine transporting fatty acids we'd only produce ATP from glucose, which supplies thirty-eight ATP molecules. Without another key nutrient, CoQ10 (the molecule that transfers electrons from Complex I and II over to Complex III), and oxygen, we'd only produce two ATP molecules via glycolysis. This perhaps gives you a glimpse into the importance of specific nutrients in the optimal functioning of mitochondria (a bit of foreshadowing, if you will). Every step of ATP production can run optimally with the proper nutrients, or in their absence, the cell is forced to produce ATP in a less efficient manner, which compromises the viability of the entire cell.

ATP Production and Turnover

While CoQ10 and L-carnitine are a powerful nutrient combination for mitochondrial health, they can't make ATP if the body doesn't have enough ADP as a raw material. After all, ADP doesn't just appear out of thin air.

Under normal circumstances, this turnover of ATP occurs millions of times in every cell each second. However, if oxygen supply is cut off or reduced, or if there is some other mitochondrial dysfunction, oxidative phosphorylation (ATP production in the mitochondria) slows or stops, causing the cell to use ATP faster than it can be replaced (remember, oxidative phosphorylation is the main producer of ATP—only two ATP molecules are produced without it).

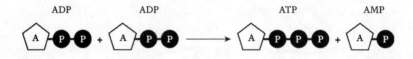

FIGURE 2.1. ATP production from combining two ADP. This is called the *adenylate kinase reaction* or *myokinase reaction,* and also results in AMP.

When this happens, ATP concentrations in the cell decrease, while the ADP concentration increases. In an effort to continue to produce ATP and normalize the ratio of ADP to ATP, the cell combines two ADP molecules to produce one ATP and one AMP (adenosine monophosphate) in a process called the *adenylate kinase reaction.*

However, while this reduces the amount of ADP building up in the cell, it increases the amount of AMP—and to very high levels when the supply of oxygen is limited. The problem here is that the cell needs to maintain a relatively constant ratio of AMP to ADP to ATP. So in order to reduce the amount of AMP (relative to ADP and ATP), the cell must degrade and eliminate this excess AMP. This reduction occurs through two biochemical pathways, and the end products ultimately exit the cell, which is not necessarily a good thing. Although the removal of excess AMP restores the AMP:ADP:ATP ratio, the absolute amount of these compounds becomes much lower—in other words, the total amount of energy the cell can produce is lower because it has lost its important building blocks. The easy analogy here is a battery and its capacity. Two AAA batteries could power the same device because they are the same physical size and have essentially the same specifications needed to power that device. However, how long they lasted could vary greatly depending on their capacity. For example, a 1,200 milliampere-hour battery has more capacity and will last longer than a 540 milliampere-hour battery, even if they are both AAA batteries.

These building blocks are called *purines*, and their loss from the cell can be devastating. Thankfully, the body will immediately start to rebuild its stock of purines, but this is a slow process and can only start with the five-carbon sugar D-ribose.

There are two biochemical pathways for D-ribose in restoring the purine pool. The first is called the *de novo* pathway, which is a very slow process. In fact, it's been calculated that it'd take over one hundred days for the human heart to make all its ATP via this pathway. The body just can't make D-ribose (the starting point for this pathway) fast enough for this pathway to be of much use in diseased states. Remember our paper recycling analogy? Creating new D-ribose from scratch is akin to planting a tree and harvesting it years later (versus recycling readily available used paper).

The second pathway is called *purine salvage*. This happens when the cell, instead of eliminating the end products of AMP degradation, retains them as building blocks to accelerate the manufacture of D-ribose. However, even with this pathway, the rate-limiting factor is again the availability of D-ribose.

The easy solution is to administer D-ribose as a supplement. In this situation, the body isn't responsible for manufacturing its own D-ribose, and so this pathway—no longer limited by availability of D-ribose—can proceed at full speed.

The importance of having a larger energy pool is that it helps to minimize the damage caused by oxygen deprivation (e.g., from reduced blood flow, called *ischemia*, such as during a heart attack or stroke). As we know, cells die when the mitochondria malfunction and cannot produce enough energy. During ischemia, oxygen levels in the cell drop and the mitochondria can no longer produce energy via oxidative phosphorylation. As the cell compensates by combining two ADP molecules to create an ATP molecule, the AMP concentration increases and then the cell needs to break this down to eliminate it. If the energy pool is low to begin with, it's depleted rapidly during ischemia. If the energy pool is large, it takes longer to deplete it from the cell. The size of the energy pool determines the extent of permanent damage to an organ experiencing ischemia, whether it is the brain, heart, or any other body part.

Restoring blood flow is absolutely the most important goal in cases of ischemia. The quicker the blood flow can be restored (and therefore, oxygen levels restored), the smaller the loss of purines from the energy pool.

It is easy to understand this need in acute cases such as a heart attack or stroke, but many diseases result in a less severe restriction of blood flow resulting in *hypoxia*, or reduced oxygen levels. Hypoxia can happen due to plaque buildup in the heart's arteries, which restricts blood flow. When blood flow is restricted, ATP is not generated as quickly, and the cell depletes its energy pool at an accelerated rate (although to a lesser degree than outright ischemia).

In these cases, the cells simply cannot supply enough energy to keep up with demand. We call these conditions by various names.

For example, when this happens to the cardiovascular system, we call it *coronary artery disease* (CAD), *angina,* or *congestive heart failure* (CHF).

Following this train of thought, I'll start my discussion on health conditions with cardiovascular disorders.

The Role of Mitochondria in Cardiovascular Disease

Cardiovascular disease is a broad category of health conditions and is likely of great interest to the majority reading this book because it's a leading cause of death globally (often alternating with cancer for the top two spots, depending on the country you're looking at). Conditions such as angina, hypertension, congestive heart failure, ischemia, and diastolic dysfunction all have their roots in mitochondrial energy. Not only can these conditions arise from a cellular energy deficiency, but they can also leak the purine building blocks of ATP out of the cell. Interestingly, when purine building blocks leak from the cell, they are metabolized to uric acid, and high uric acid in patients is often reflective of dysfunctional ATP metabolism (an important point to understand for clinicians treating gout, for example).

It can take up to two weeks (and in some cases months) for the heart to produce enough ATP, by natural, built-in mechanisms, to offset the deficit caused by ischemia. Also, since the heart is constantly consuming ATP, it's difficult to make up for this energy deficit quickly, and most patients with ischemia will need to take targeted nutritional therapy to help restore the energy balance. I'll discuss these nutritional therapies in depth in chapter 3.

Understanding Smooth Muscles

A large part of our cardiovascular system involves smooth muscles (muscles not under voluntary control), so let's review their significance and what their normal and abnormal functioning look like. Smooth muscles are found in the blood vessels of the cardiovascular system, but are also contained within other organs and tubes in the body, including the stomach, intestines, bladder, airways, uterus, and the penile and clitoral cavernosal sinuses. Bundles of smooth muscle cells are also attached to the hairs of the skin and to the iris and lens of the eye.

Smooth muscle cells receive input from the autonomic nervous system (which is the part of the nervous system not under voluntary control, e.g., the part that digests food). In addition to the autonomic nervous system, smooth muscles are controlled by hormones and other local chemical signals. Smooth muscle cells also develop tonic and phasic contractions in response to changes in load or length. By contrast, skeletal muscles are under voluntary control, and these are the muscles we consciously contract and relax when we decide to move our arms or go for a walk.

Contraction (shortening of muscle cells) in smooth muscles is a highly regulated process. In some smooth muscle cells, the contraction is maintained at a low level in the absence of external stimuli. This activity results in what is known as *smooth muscle tone* and its intensity can be varied. Keep this in mind when I discuss how this relates to conditions such as hypertension (see "Coenzyme Q10" on page 144).

Regardless of the stimulus, a smooth muscle contraction is initiated by calcium ions entering the cytosol (from the sarcoplasmic reticulum—a membrane-bound structure in muscle cells that stores calcium) and binding to a calcium-binding messenger protein called *calmodulin*. This stimulates another protein called *myosin* (the protein that contracts and is dependent on ATP) to attach to actin in *cross-bridge cycling*.

Initiation of relaxation, on the other hand, begins with the removal of calcium ions from the cytosol and stimulation of an enzyme that deactivates myosin (referred to as *myosin phosphatase*).

The Importance of Smooth Muscle Relaxation

Many people don't realize that muscle relaxation (elongation of a muscle cell) requires considerable amounts of energy. Whether it is a conscious decision to relax skeletal muscles or the involuntary relaxation of smooth muscles, the process requires a decreased concentration of calcium ions. All this calcium must move out of the cytosol and into the sarcoplasmic reticulum. However, this process requires the use of a pump because the calcium must move *up* the concentration gradient—and going against the gradient requires energy. That energy, of course, comes from ATP. The enzyme embedded in the membrane of the sarcoplasmic reticulum, called *calcium-magnesium-ATPase* (Ca-Mg-ATPase), when activated, binds two calcium ions, which are then transferred to the inner part of the sarcoplasmic reticulum and released (sequestered, ready for the next stimulus signaling a contraction).

This pump also has two ATP-binding sites, and both sites must have ATP attached for it to work. However, there are intricacies. The first ATP-binding site has a high affinity for ATP, and therefore, any ATP in the vicinity binds to this site readily. Once bound to this site, ATP releases its energy, and is turned into ADP. The second ATP-binding site does not attract ATP so easily. In fact, the only way for ATP to bind to the second site is to ensure a high concentration of ATP so that hopefully one will just "fall" into the binding site. Building up this concentration obviously requires significant amounts of ATP to be produced.

The state of rigor mortis, when our muscles become tense and rigid after death, is a good example of how relaxation requires more ATP than contraction. In death, fuel and oxygen are no longer delivered to the muscles, and ATP production stops. Without enough ATP, the calcium ions cannot be pumped out of the cell, and the muscles can no longer "relax."

Magnesium ions are also necessary for the activity of the Ca-Mg-ATPase; they bind to the catalytic site of this enzyme to mediate the reaction. Without magnesium, this enzyme cannot function and relaxation of the smooth muscle cannot occur (which can lead to things like high blood pressure, heart problems, or restricted breathing). For those who might have heard magnesium is great for muscle function and relaxation but didn't know how or why, now you know.

The Basics of Cardiac Physiology

Now let's discuss the other part of the cardiovascular system: the heart itself. The human heart has four chambers—two upper chambers (called the *left* and *right atria*) and two lower chambers (called the *left* and *right ventricles*). In terms of heart function, *systole* describes the stage of a heartbeat when the ventricles contract, squeezing the blood out to the arteries. This contraction ejects most of the blood out of the ventricles, and the percent of blood that's pumped out (relative to its starting point when "relaxed") is called the ejection fraction (normal range is 50–70 percent). While it's easy to see how contraction of the heart requires energy, this stage requires the least amount of energy in the cycle of a heartbeat. Muscle cells in general (not just smooth muscle cells), including the heart, are able to contract even when energy levels are extremely low (they just might not be able to relax again).

After this systolic phase is the *diastole*, or "relaxation," phase when the ventricles fill up with blood. The diastole phase generally lasts less than a third of a second but requires the most ATP. There are two reasons, both of which were just discussed. First, energy is required to separate the bonds formed during the contraction phase, which allows the muscle to return to its relaxed state. Second, the removal of calcium ions from the cell requires energy.

Without enough ATP, the calcium ions cannot be pumped out of the heart muscle cells, and the heart can no longer relax and fill up with blood efficiently. This is called *diastolic dysfunction*. The beginning stages of diastolic dysfunction are characterized by a thickening (called *hypertrophy*, or enlarging of the heart muscle, usually specific to the left ventricle) and stiffening of the ventricular walls. The combination of hypertrophy and stiffening causes blood pressure to rise, reduces the amount of blood that's pumped out per contraction (lower ejection fraction), and makes it more difficult for the heart to relax and fill up properly (which propagates this progressive cycle).

Despite seemingly normal systolic function, diastolic dysfunction is an early sign of serious heart problems around the corner—namely, congestive heart failure. Preserving diastolic function in patients is a major goal for cardiologists, and the solution is to ensure an abundant pool of ATP energy.

Another energy-intensive process in a heartbeat is maintaining a proper ionic balance. The proper flow of ions in and out of a heart muscle cell is essential to maintain the normal electrochemical gradient across the cell membrane. This gradient is what's responsible for maintaining regular heart rhythm. When this gradient is disrupted, the result is irregular or "skipped" heartbeats (*arrhythmia*) or another type of abnormal contraction or rhythm.

All of these high energy demands must be met by a small pool of ATP. As a result, the ATP supply must be continuously replenished—and again, that is the job of our mitochondria. I'm hoping that by now you can appreciate the important role mitochondria have in one of the leading causes of death; but you'll soon see that their importance goes far beyond the cardiovascular system, and plays a crucial role in essentially every functional system in the body.

The Role of Mitochondria in the Nervous System, Brain, and Cognitive Health

Tissues with a high demand for energy are uniquely dependent on the energy delivered by mitochondria and, therefore, also have the lowest threshold for displaying symptoms of mitochondrial dysfunction. Thus, the central nervous system is often one of the first systems to display outright symptoms of bioenergetic deficiencies. Large amounts of energy are required by neurons (nerve cells) to carry out their specialized functions.

In fact, while the brain makes up only about 2 percent of a person's body weight (which of course varies depending on the person), at rest, it consumes about 20 percent of the total energy the body needs. The brain is a giant tangle of countless neurons, so it would stand to reason that this organ might suffer significantly from mitochondrial dysfunction, and possibly respond best to mitochondrial nutrients.

Stroke: Suffocating the Brain's Mitochondria

The cycling of blood through the circulatory system delivers a constant supply of oxygen, glucose, and nutrients to every living cell. The brain consumes a disproportionate share of the body's circulating blood flow (14 percent) and oxygen (20 percent), yet despite its extraordinary energy demand, the brain's energy reserves are actually very small. The brain's metabolism can sustain energy production for only about one minute before it needs to be replenished. Thus, nerve cells are particularly vulnerable to ischemia (reduced blood flow) and hypoxia (reduced oxygen levels).

When the steady flow of blood through a portion of brain tissue ceases—as from a clot or hemorrhage—metabolism rapidly fails in the brain cells. As the oxygen supply runs out, the cells shift to anaerobic metabolism for the short term, but after a few minutes without blood, neurons suffer irreversible injury.

During a stroke, however, blood supply is not cut off uniformly. Instead, impairment in circulation is more severe toward the core of the affected area, where the lack of flow might be almost complete. Cells in the core tend to die quickly through *necrosis*. Remember, as opposed to *apoptosis* (programmed cell death that happens in an orderly fashion), necrosis is messy. Cells break apart, spilling their contents into nearby tissue, which further aggravates the situation by causing inflammation.

An important consequence of a stroke is that cells in the surrounding areas die off hours or days later. How and why this happens remains a mystery; however, this secondary brain damage is considered potentially preventable. A growing body of research suggests that the focus of both primary and secondary stroke damage—and potential intervention or therapy—lies in the mitochondria.

What's interesting is that mitochondria might actually suffer greater injury when blood flow is reduced rather than when it stops altogether. Complete blockage of blood flow also cuts off the supply of oxygen, which in turn reduces oxidative stress and free-radical production. When blood flow is merely *reduced*, a small amount of oxygen continues to flow, generating further free radicals in addition to those resulting from stroke-impaired cellular respiration. These additional free radicals could also explain the secondary damage that occurs days later.

A paper published as early as 1992 suggested that the transition from reversible to irreversible ischemia depends on the functional state of the mitochondria. More specifically, their ability to restore oxidative phosphorylation determines functional recovery.

With their high sensitivity to reduced blood flow, brain mitochondria exhibit the first signs of injury even during a moderate reduction of cerebral blood flow. Injury to mitochondria during and after a stroke has many consequences, and includes impaired bioenergetics, free-radical damage, calcium dysregulation, increased excitotoxicity, and promotion of programmed cell death. Unfortunately, as typical for mitochondrial impairment, mitochondrial brain injury causes further mitochondrial damage in those brain cells lacking blood flow, generating a vicious cycle of cellular injury after a stroke.

Regardless, the primary goal is to restore blood flow and oxygen levels in the brain. Makes sense, right? Well, not so fast. Restoring blood flow *is* the primary goal, but this can quickly turn into a very dangerous situation. When blood flow—and oxygen—is restored to the affected area, there is even *greater* injury to the mitochondria, called *ischemia-reperfusion injury* (IRI, or simply, *reperfusion injury*). IRI is also frequently seen after cardiac surgery. In some cases, such as a planned cardiac surgery, we can prepare for IRI and take measures to reduce this damage by priming the cells and mitochondria for this added stress through interventions such as nutritional therapy. In most cases, such as a heart attack or stroke, there's no advance warning, so no protective preparations are in place. Let me explain how this all unfolds.

With reduced blood flow in the brain, ATP is still being used as usual for regular cellular activity, but is not being produced, and so the cell combines two ADP molecules to form ATP. As discussed earlier, this results in an increasing concentration of AMP, which is then eliminated from the cell, reducing the overall energy pool of the cell. Further, the absence of oxygen and fuel creates a condition where the mitochondria go into a low-use state, much like hibernation. When blood flow is restored, we see a perfect storm of events: a rush of fuel and lots of oxygen to the brain, a lag period for the mitochondria to "wake up," and a deficiency of the building blocks of ATP (they were eliminated from the cell as AMP). So even after the mitochondria wake up and are running at full speed, there is still an ATP deficiency because the purine nucleotides have been lost (which means ADP/ATP recycling is reduced). The result is a very high rate of free-radical production, rather than restoration of normal function. This high rate of free-radical production pushes any cells already close to their threshold for irreversible damage or death past that line, resulting in the damage that's typically seen with IRI.

The extent of the neurological damage caused by a stroke depends on this secondary brain damage that results from IRI. The question being debated in the field of medical research is whether this delayed death of neurons occurs through necrosis or apoptosis—or something in between. Much of the recent evidence suggests apoptosis as the major factor in delayed neuronal death. While this cellular suicide removes damaged cells in a neat, orderly way, a stroke might trigger it accidentally in healthy cells. Whatever the mode of cellular death, minimizing delayed neuronal death is critical in preserving brain function to the greatest extent possible. By nurturing our mitochondria, which is what I'll discuss in chapter 3, we can prep them to have the greatest chance of survival in these types of events.

Mitochondrial Involvement in Neurodegeneration

As early as 1999, a series of published scientific reviews summarized the mounting body of evidence on the role of the mitochondria in neurodegeneration. One review by Cassarino and colleagues stated, "It is becoming clear that subtle functional alterations in these essential cellular dynamos can lead to insidious pathological changes in neurons." The authors outlined a theory of neurodegeneration based upon a vicious cycle of mtDNA mutation, bioenergetic decline, and free-radical damage—the same story now seen in a host of other disorders and confirmed by further studies over the last twenty years or so.

These published studies now corroborate the role of abnormal mitochondrial dynamics in neuronal cell death and the onset of Alzheimer's, Parkinson's, and Huntington's disease and other neurodegenerative disorders. Although many health conditions, aging, and general neurodegeneration have similar basic causes, the physiology of the brain is unique in certain ways, and its pathologies present some interesting mechanisms and features.

The brain is particularly vulnerable to free-radical damage (due to its rich oxygen supply and high fatty acid content), so it might seem logical to assume the brain's antioxidant defense system is especially powerful. Unfortunately, it is not, and this delicate organ is relatively under-defended against free-radical damage. As a result, the cells of the brain gradually accumulate oxidative damage over time. This is true for everyone, but is of particular concern for those with a genetic or environmental predisposition to neurological degeneration.

Most of the brain's fatty acid content is contained in the cell membranes, their extensions (such as axons and dendrites), and their mitochondria. As we age, more of these lipids become oxidized from being exposed to the brain's rich supply of oxygen and free radicals, and the brain's vulnerability to degenerative diseases increases. Maintaining mitochondrial health is an important strategy in preventing this slow decline in our mental faculties as we age.

Excitotoxicity

In the late 1980s, scientists at the National Institutes of Health proposed that *excitotoxicity* (toxicity from overstimulation of nerve cells) develops when the energy level of neurons declines. Subsequent research has confirmed this, as studies have shown that CoQ10 (remember, this is the compound that shuttles electrons from Complex I or II to Complex III, and can be found as a nutritional supplement) protects against excitotoxicity by raising energy levels in nerve cells.

The neurotransmitter *glutamate* normally transmits excitatory impulses. In neurodegeneration, however, the brain becomes chronically oversensitive to glutamate, which then becomes a slow-acting "excitatory toxin" on brain cells. For mitochondria, this means that they are constantly under the direction to produce more energy— more energy than the neurons actually need. With this higher rate of activity comes a higher rate of free-radical production and, over time, comes the accelerated demise of these mitochondria. Ultimately, this chain of events results in dysfunction within the neurons.

Mitochondria in Neuronal Signaling

Brain cells talk to one another in a variety of strengths or intensities. Sometimes they speak loud and clear, but other times they whisper or mumble. For years scientists questioned why and how neurons change their intensity so frequently. A study by Sun and colleagues published in the summer of 2013 showed that rapidly moving mito-chondria emit bursts of energy, and this might regulate neuronal communication.

The network of neurons throughout the body controls thoughts, movements, and all senses by sending and receiving thousands of neurotransmitters (the brain chemicals) at communication points between the cells called *synapses*. These neurotransmitters are sent from tiny protrusions found on the neurons, called *presynaptic boutons*, which are aligned along the axons. They help control the strength of the signals sent by regulating the amount of neurotrans-mitters released as well as the manner in which they are released.

The production of neurotransmitters, their packaging and release, and the reception or removal of these chemicals all require energy. Previous studies showed that mitochondria can move rapidly along axons, dancing, in a sense, from one bouton to another. This study showed that these moving mitochondria might control the strength of the signals sent from boutons. The researchers used advanced techniques to watch mitochondria move between boutons while they released neurotransmitters, and found that boutons only sent consis-tently strong signals when mitochondria were nearby—when the mitochondria were absent or moving away from boutons, the signal strength fluctuated. These results suggest that the presence of stationary mitochondria at synapses improves the stability and strength of the nerve signals. This is the type of data we need to confirm mitochondrial involvement in the pathogenesis of neurodegenerative diseases and any neurological disease that requires efficient and appropriate transmis-sion of nervous system signals (such as depression, ADHD, and so on).

To test this further, the researchers manipulated mitochondrial movement by changing levels of *syntaphilin*, a protein that helps anchor mitochondria to the cell's cytoskeleton inside the axons. Removal of syntaphilin resulted in mitochondria that moved quicker, and the electrical recordings from these neurons showed that the signals they sent fluctuated greatly. On the flip side, elevating syntaphilin levels slowed mitochondrial movement and resulted in boutons that sent signals with the same strength. Previously, it had been shown that about one-third of all mitochondria in axons move about; the rest are stationary. Nerve cell communication is obviously tightly controlled by highly dynamic events occurring at numerous synapses.

The researchers also found that blocking ATP production in mitochondria reduced the strength of the signals sent even if mitochondria were near the boutons. Problems with mitochondrial energy production and movement throughout neurons have been implicated in Alzheimer's, Parkinson's, ALS, and other major neuro-degenerative diseases; this 2013 research adds a key piece of the puzzle and gives us more reason to target mitochondria and cellular energy in these illnesses.

Alzheimer's Disease: Don't Forget the Mitochondria!

Alzheimer's disease is the most common form of adult-onset dementia; an eighty-year-old has about a 30 percent chance of developing Alzheimer's. Its pathology is complex and while we cannot yet clearly separate causes from effects, recent research has discovered many of the mechanisms involved in the disease.

At the cellular level, there is an extensive loss of neurons and there are high levels of insoluble fibrous deposits (known as *senile plaques* and *neurofibrillary tangles*). At the core of the plaques is a toxic protein called *amyloid beta*—the hallmark of Alzheimer's—that attacks cells on several fronts. Amyloid beta generates free radicals, damages mtDNA, impairs cellular bioenergetics, and alters the proper folding of proteins so that they form neurofibrillary tangles. However, there is evidence to suggest that the formation of amyloid beta is the brain's way of defending *against* oxidative stress (a result of Alzheimer's, not a cause). In her recent book, *The Alzheimer's Antidote*, Amy Berger does an excellent job of assessing the current state of affairs and offers diet and lifestyle-related solutions based on the newest information and research. Despite some seemingly contradictory studies and insights, Berger's conclusion is that Alzheimer's is a metabolic disorder. Based on the research I've done for this book, I would agree that the long-held beliefs about Alzheimer's are falling apart and the most current research shows we need a greater focus on mitochondria in the prevention and treatment of the disease.

According to the results of some studies, the degree of disability in Alzheimer's disease correlates with the level of bioenergetic impairment in the brain. In fact, a recent study suggests that cellular energy production might be a better indicator of Alzheimer's disease severity than senile plaques. In this particular study, degree of clinical disability did not correlate with the density of senile plaques, but did correlate with a mitochondrial abnormality involved in cellular energetics.

Regardless of whether amyloid beta is a cause of or result of oxidative stress, another potent free radical called *peroxynitrite* (formed from nitric oxide) oxidizes lipids in the membranes of nerve cells. This generates the highly toxic by-product *hydroxynonenal* (HNE), which is found in excess quantities in multiple brain regions of Alzheimer's patients. HNE kills brain cells not only directly but also indirectly by making them more susceptible to excitotoxicity. (As a quick side note, CoQ10 and vitamin E can protect cell membranes from lipid peroxidation, and CoQ10 has been found to reduce peroxynitrite damage and HNE formation in the bloodstream.)

Current research has not identified a single underlying cause of Alzheimer's disease, but an interesting multiple-factor theory was proposed by Wan-Tao Ying in 1997. According to this theory, Alzheimer's disease develops from the interplay of four causes: imbalances in APP (amyloid precursor protein), calcium, free-radical damage, and bioenergetic deficit. Ying cites studies showing that each factor reinforces, and is reinforced by, each of the other factors.

Overeating and Alzheimer's Disease

Data presented in 2012 by the Mayo Clinic Study on Aging suggested that consuming between 2,100 to 6,000 calories per day might double the risk for mild cognitive impairment (MCI, the precursor of Alzheimer's) in adults aged seventy years and older (compared to those who consumed less than about 1,500 calories daily).

While previous studies noted an association between different dietary patterns and risk for MCI (for example, elderly individuals who follow a healthy Mediterranean diet are less likely to develop MCI, and are also less likely to progress from MCI to Alzheimer's disease), this was the first study to find the association between excessive daily calorie consumption and MCI. All of these studies together suggest that we need to ensure we're eating a nutrient-dense diet, and minimize empty calories that come from foods such as added sugar, white bread, and salty snacks.

Excess calories have also been linked to countless other degenerative diseases. On the flip side, as previously discussed, caloric restriction (a reduction in calories while ensuring nutrient needs are met) has been linked to life extension and possibly a reduction in the risk of degenerative diseases. This adds weight to the free-radical and bioenergetic components of Ying's multiple-factor theory. (I'll discuss caloric restriction in further detail in chapter 3, "Ketogenic Diets and Calorie Restriction," page 170.)

There is a significant amount of current research in the area of Alzheimer's disease, and I hope we will soon have greater clarity over how we can help those who have this condition.

Parkinson's Disease: Rethinking L-Dopa Therapy

Recent research in animal models of Parkinson's disease suggests that CoQ10 can protect brain cells from neurotoxicity and excitotoxicity, even in cases where other powerful antioxidants cannot. This finding is significant because it draws focus to the importance of mitochondrial dysfunction and cellular energy in Parkinson's disease. Subsequent research has further confirmed that dysfunctional cellular energy production plays a large role in the progression of Parkinson's disease.

In Parkinson's disease, cell death is primarily targeted to the neurons in the *substantia nigra*—a part of the brain that coordinates movement. These neurons produce the neurotransmitter *dopamine*; the death of these cells depletes dopamine stores, and ultimately leads to muscle rigidity, tremors, and difficulty initiating movement.

Research has shown that the substantia nigra is the part of the brain that has the greatest number of mutations in mtDNA, and human evidence reveals that the mitochondria of patients with Parkinson's disease exhibit several deficiencies. One of the most well-characterized deficiencies is diminished Complex I activity (the first complex in the ETC). In rat studies, Complex I inhibition has been observed to directly follow the administration of dopa or dopamine in a dose-dependent manner. Other rat studies have shown a dose-dependent increase in hydroxyl free radicals in the mitochondria when administered dopa. These studies were the first indication that perhaps increasing the amount of a substance that we think is deficient might not be the answer.

As previously discussed, superoxide radicals are generated when electrons leak and react with oxygen. Deficits in Complex I increase leakage of electrons and, in turn, increase superoxide production, and ultimately diminish ATP production (you might remember that Complex I is the primary site for superoxide radicals). As superoxide is neutralized, hydrogen peroxide is generated in the interim. As hydrogen peroxide is broken down, hydroxyl radicals can be produced instead of water. This is consistent with the observation that hydroxyl radical production is increased when Complex I is inhibited. The question is, why are hydroxyl radicals formed instead of water? The answer relates to iron in its reduced form (Fe^{2+}), which catalyzes the breakdown of hydrogen peroxide into hydroxyl radicals. For this reason, the association between tissue iron stores and Parkinson's disease (incidence and progression) should receive more attention.

Further, as alluded to earlier, research into mitochondrial dysfunction in Parkinson's disease has raised serious questions regarding the conventional use of L-dopa in the treatment of the disorder. L-dopa is prescribed in conventional medicine for its ability to improve the symptoms of Parkinson's (at least temporarily), but it does not improve the underlying disease pathology. There is growing evidence that L-dopa might actually *aggravate* some of the underlying causes of Parkinson's disease. As such, it might be time to reconsider the costs and benefits of L-dopa therapy. In fact, it's well known that L-dopa therapy eventually loses its effect, and the symptoms return with vigor. Is short-term relief of symptoms worth accelerating the progression and increasing the severity of the disease?

The mitochondria of Parkinson's patients also exhibit some (although milder) inhibition of Complex III activity (the second most prominent site of superoxide generation).

A relative deficit of *alpha-ketoglutarate dehydrogenase complex* (KGDHC, a key enzyme of the TCA cycle found in the mitochondrial matrix) has also been noticed. KGDHC produces NADH, the substrate for Complex I, and has been found to be significantly depleted in the lateral part of the substantia nigra regions of Parkinson's patients. Interestingly, reductions in KGDHC levels have been noted in the cortex of Alzheimer's patients as well.

M. Flint Beal, a prominent neurologist, has spent years proving that CoQ10 has neuroprotective properties that might help diseases such as Parkinson's and Huntington's, and a growing body of scientific evidence supports his hypothesis. Research has established that mitochondria from platelets in patients with early, untreated Parkinson's had reduced activities of Complexes I, II, and III (compared to age-matched controls).

His research team also demonstrated that CoQ10 administration in middle-aged and old-aged rats could restore levels of the nutrient to those of younger rats. The results showed that CoQ10 levels rose by 10–40 percent in the mitochondria of the cerebral cortex region of the brain. In a later study of mice, oral supplementation with CoQ10 attenuated chemically induced neurotoxicity (which had been shown to cause a Parkinsonian syndrome in test animals). After a number of weeks of exposure to the Parkinson's-inducing chemical, striatal dopamine concentrations and dopaminergic axon density were reduced, but they were much higher (37 percent and 62 percent, respectively) in the group that was pretreated with CoQ10, confirming a bioenergetic deficit is a component of the disease.

Depression

Up to 20 percent of the population will experience some form of a stress-associated disease, including depression, yet despite decades of research, we still do not completely understand this complex brain disorder. There is controversy about considering depression a physical illness, in part because a reproducible, sensitive, and specific biologic marker is not available. However, there is evidence that mitochondrial dysfunction and free radicals might be associated with abnormal brain function and various mood disorders, such as depression. Evaluation of mitochondrial dysfunction in specific tissues could broaden the perspective of depression beyond theories about neurotransmitters or receptor sites, and could explain the persistent signs and symptoms of depression.

An increasing body of evidence implicates mitochondria in the etiology of depression. It has been proposed that the balance between stress response (the adaptation to our constantly changing environment) and available energy (our mitochondrial function) is crucial for mental health. More specifically, stress activates different areas of the brain, and changes its structure and function (what is called *neuroplasticity*). This comes at a metabolic cost, and of course, it's the mitochondria that are responsible for meeting this additional energy demand.

People with optimal mitochondrial function can cope with the energy demands of stress-induced neuroplasticity, which means that these individuals are at relatively low risk for depression. In individuals with mitochondrial dysfunction, on the other hand, stress-induced depletion of the brain's energy supply could ultimately compromise neuroplasticity, which, in time, could render an individual vulnerable to clinical depression as the adaptation response falters.

I don't mean to imply that all mitochondrial disease patients suffer from depression, or that all depressed patients have underlying mitochondrial dysfunction. However, mitochondrial dysfunction could be a cause of depression in a subgroup of patients. If so, this will have a profound effect not only on our understanding of depression but also on its treatment.

Attention-Deficit/Hyperactivity Disorder: Pay Attention to the Mitochondria

Attention-deficit/hyperactivity disorder (ADHD) is a heterogeneous condition that affects a significant, and growing, proportion of the population. ADHD is defined by persisting, developmentally inappropriate, cross-situational, and impairing levels of inattention, impulsiveness, and hyperactivity. Several studies have linked ADHD with markers of elevated oxidative stress and free-radical damage, and increasing evidence also documents that ADHD is linked to environmental pollutants that are known to adversely affect mitochondria.

Whether induced by pollutants, illness, or predisposing genetics, elevated free-radical damage and its effects upon mitochondria might be a significant factor in some—and potentially many—cases of ADHD. This indicates that treating mitochondrial dysfunction might be beneficial for some individuals who have ADHD.

As we already know, the mitochondrial dysfunction that arises from free-radical damage will result in an energy deficiency, which leads to inefficient and inconsistent neuronal transmission of information by the *astrocyte* (the major non-neuronal component of the central nervous system). Astrocytes play an important role in providing energy by supplying lactate to rapidly firing neurons. Astrocytes can also provide lactate to *oligodendrocytes*, where lactate is used as a substrate for *myelin synthesis*, and thus enables rapid *neurotransmission* (myelin allows nerve signals to travel quickly, which is normal). The presence of receptors for the major brain neurotransmitters on astrocytes adds to the robust evidence for their direct involvement in neurotransmission.

According to one bioenergetic theory called the Energy-Deficiency Model, the genesis of the behavioral symptoms in ADHD is directly linked to impairments in the astrocyte-neuron lactate shuttle. This shuttle is based on the astrocytes' uptake of glucose from the blood, its storage as glycogen, and its conversion to lactate.

This is critical, because as I mentioned, the brain at rest uses about 20 percent of the body's energy. Its energy demands are massive. Neural activity triggers the uptake of glucose from the blood, into the astrocytes. It also triggers the astrocytes to break down stored glycogen into glucose. This glucose (from both the blood and stored glycogen) gets metabolized to lactate, which in turn gets shuttled over to the neurons. The neurons then take this lactate into the TCA cycle and then on to oxidative phosphorylation to produce ATP.

Lactate is the essential energy source for rapidly firing neurons, and is a more efficient fuel source than glucose. Why? Because it is metabolized to form ATP more rapidly (glycolysis has already occurred) and, unlike glucose, it does not require ATP for its metabolism (remember: it takes two ATP molecules to convert glucose to lactate). Due to their high energy demands, it is imperative for neurons to make use of the most efficient energy supplies when rapid neural processing is required, and the brain has evolved ways to do so.

However, in ADHD, this lactate production by astrocytes is not sufficient to supply rapidly firing neurons with energy during brief periods of increased demand. The insufficient amount of lactate leads to a localized and transient deficiency in ATP production, impaired restoration of ionic gradients across neuronal membranes (which requires energy to restore ions against a concentration gradient), and slowed neuronal firing. The consequence of all this is inconsistent performance of demanding cognitive tasks.

These periods of rapid firing are then followed by slow unsynchronized firing, which exerts less demand on energy resources. This allows replenishment of energy reserves and restoration of function. The brief periods of energy deficiency, followed by periods of normal supply, are suggested to account for the variability of behavioral response seen in patients diagnosed with ADHD when performing complex tasks that require speed and accuracy.

Glutamate, a major excitatory neurotransmitter, stimulates *glycolysis* (glucose utilization and lactate production) in astrocytes. However, there is a downside to glutamate, as anyone sensitive to MSG (monosodium glutamate) would know. While glutamate is excitatory to the neurons, long-term exposure quickly depletes energy stores. Many who are sensitive to MSG (and not just those with ADHD) report a racing heart and sweating (referred to as *excitation*) followed by extreme fatigue (energy depletion). Astrocytes normally maintain low extracellular levels of glutamate, and this is achieved with the help of the membrane potential (the electrochemical gradient across the neuronal membrane). However, due to the impaired ability to maintain electrochemical gradient in ADHD, removal of glutamate from the extracellular fluid is hampered. Failure to maintain low levels of extracellular glutamate not only impairs glutamate's neurotransmitter function, but also affects neuroplasticity, learning, and memory. Cell death can also result from this overexcitation (which means the mitochondria are pushed to their limit in energy production, which subsequently results in excessive free-radical damage, and the chain of events that leads to apoptosis). ADHD patients are known to have reduced gray matter in their brain; cell death due to mitochondrial dysfunction eventually results in the atrophy of any affected organ.

This process also helps explain why the conventional medical field prescribes amphetamines to patients who have ADHD. *Methylphenidate* (Ritalin), for example, has been shown to increase glucose utilization by astrocytes in brain regions. This drug helps ADHD patients who have decreased glucose utilization and impaired energy supply, and might also help others who have impaired myelination from a longer-term lactate deficiency. Intrinsically, however, most would agree that administering amphetamines to children just doesn't sit well in the psyche—especially when schools consistently teach children how bad drugs are; it sends mixed and confusing messages. There is a better way, which I'll discuss in chapter 3.

Chronic Fatigue Syndrome, Myalgic Encephalomyelitis, and Fibromyalgia

Although *chronic fatigue syndrome* (CFS), *myalgic encephalomyelitis* (ME), and *fibromyalgia* are all distinct and separate conditions, due to their similarity and significant overlap in symptomatology, they are often discussed together. For the purposes of this book, I'll also group them together. In terms of diagnosing these disorders, however, pain is the main factor for fibromyalgia—specifically, pain and tenderness in certain areas of the body when pressure is applied to them (these are called *tender points*). Most of the other symptoms (such as fatigue, cognitive dysfunction, headaches, and sleep disturbances) are common to all three.

We know by now that problems arise when the production of cellular energy is stressed. Unfortunately, most people with CFS are constantly in a state where demand outstrips supply. Investigations by Dr. Sarah Myhill and colleagues who have studied various possible root causes of CFS have concluded that CFS is a result of mitochondrial dysfunction.

In a state of constant undersupply of energy, some ADP is combined together to form ATP, but this combination also creates AMP—as discussed earlier. Remember, AMP is eventually eliminated from the cell, which means a critical building block of ATP is lost. In order to rapidly meet the energy demands of the cell in the presence of a reduced energy pool, the body will make very small amounts of ATP directly from glucose through glycolysis (or *anaerobic metabolism*), which is much faster in generating energy, but far less efficient than aerobic metabolism. Switching to this anaerobic state is exactly what many people with CFS do.

Unfortunately, as Myhill points out, this switch results in two serious problems. First, lactic acid (the product of anaerobic metabolism) quickly builds up—as it also does in someone who is sprinting, where energy production must be fast—resulting in aches and pains known as the "lactic acid burn." The person switches to anaerobic metabolism to meet the high energy demands in the short term, but pays for it later with this lactic acid buildup. Second, using glucose in this way means very little, if any, is available to make D-ribose. So people with CFS are never fully able to recover and get ahead in the progress of rebuilding their energy pool and capacity.

For those wanting to dive deeper into CFS, I encourage you to read more on Myhill's website, which is full of great information, in addition to her book, *Diagnosis and Treatment of Chronic Fatigue Syndrome and Myalgic Encephalitis*. It's fascinating to learn that when you follow the trail of evidence, mitochondrial dysfunction explains not only the muscle soreness and tenderness and intense ongoing fatigue but also how and why people who have CFS experience poor cardiovascular function, why they can't tolerate heat yet don't sweat as they should, why digestion is a mess, and why brain fog is a common complaint. One domino knocking down the next, it all starts with mitochondrial dysfunction.

Type 2 Diabetes

Commonly referred to as just diabetes, this metabolic disorder is characterized by high blood sugar over a prolonged period of time. Previously, a diagnosis of diabetes meant a death sentence, but with advancing medical research, it's no longer the scary disease it used to be. Yet it's important to remember that if left untreated, diabetes can lead to many complications, including cardiovascular diseases, neurological disorders, stroke, kidney failure, and even coma.

There are two main types of diabetes, type 1 and type 2. Type 1 occurs when the body's immune system inappropriately attacks and destroys the insulin-producing cells in the pancreas. Type 1 is an *autoimmune disease*, and as a result of this destruction, the lack of insulin can no longer keep blood sugar levels in check. Type 2, on the other hand, occurs when the body can't effectively utilize the insulin that is produced and released. Most of our discussion here will focus on type 2, which also happens to account for about 95 percent of the cases of diabetes.

While the management of diabetes has come a long way, recent evidence suggesting mitochondrial dysfunction lies at the heart of this disorder has given us insight into how to stop its progression, and even reverse it! The reversal of diabetes, which I'll discuss in chapter 3 (see "Ketogenic Diets and Calorie Restriction" on page 170), is our generation's best example of how focusing on mitochondrial function and health can have tremendous benefits to what was previously thought to be an irreversible disorder.

Mitochondrial Damage in Type 2 Diabetes

Recently, mitochondria have been identified as playing an important role in pathogenesis of type 2 diabetes, which is associated with both impaired insulin release and impaired insulin action at target tissues. Evidence suggests that mitochondria play a role in both processes, and defects are evident early in the course of the disorder.

On the flip side, type 1 diabetes results mainly from the autoimmune destruction of beta cells in the pancreas. Mitochondria also are important in this situation—possibly in the pathogenesis, but certainly in the treatment and prevention of long-term consequences of type 1 diabetes.

Regardless of the cause, both types of diabetes (and less common forms, such as mitochondrial diabetes, discussed in the next section) are associated with similar long-term complications because they appear to result from pathogenic processes at the mitochondrial level.

Mitochondrial function has different implications for diabetes in different cells and tissues, but at this point, the scientific community places emphasis on the major cell types responsible for insulin secretion (pancreatic beta cells), the targets of insulin's action (skeletal and cardiac muscle cells, and liver cells), as well as target organs for the major complications of diabetes (kidneys, retina, nerves, and vascular cells).

Type 2 diabetes is well known to be a progressive disorder, and the general consensus is that a target cell's insulin sensitivity is substantially impaired (referred to as *insulin resistance*) early in the course of the disease, whereas worsening of blood sugar over time is related to beta-cell dysfunction. The result is a diminished ability to secrete enough insulin to keep up with the demand imposed by insulin resistance.

Inability to keep up with the demand in turn leads to *hyperglycemia* (or high blood sugar), which is a major medical concern. Research has shown that hyperglycemia induces mitochondrial superoxide production in the *endothelial cells* (the cells that line the blood vessels), which is an important mediator of diabetic complications such as cardiovascular diseases. Endothelial superoxide production also contributes to atherosclerosis, hypertension, heart failure, aging, and sepsis.

Further, the high glucose levels seen in diabetes will "glycate" proteins, known as *AGEs* (advanced glycated end products). These proteins have altered functions, from simply no longer working (which is still quite dangerous, even though it sounds benign) to, worse, harming the cell's functioning. These glycated proteins can also bind to mitochondria and compromise their function.

In addition, the skeletal muscles of people with type 2 diabetes have shown a reduced capacity of the ETC, and the mitochondria are smaller than normal. Mitochondrial damage also appears to be a major cause of lipid accumulation in these cells. PPARG coactivator 1 (PGC1) is a key factor located in the mitochondrial matrix for lipid oxidation, and the expression of PGC1 is reduced in type 2 diabetes patients. The accumulated lipids turn into cytotoxic compounds, damaging the mitochondria and leading to insulin resistance.

Lipotoxicity is when accumulated lipids lead to functional impairment of certain cells. Compared with other types of lipids, free fatty acids are more toxic to the cells. Several researchers have identified that lipotoxicity and lipid accumulation can promote the progression of type 2 diabetes. Defects in the capacity to metabolize fatty acids in skeletal muscles are a common characteristic of type 2 diabetes. Under normal physiological conditions, lipids are metabolized through beta-oxidation in mitochondria. However, with mitochondrial damage, lipids cannot be metabolized normally and so fatty acids accumulate.

Fatty acids are particularly prone to oxidative damage, resulting in the formation of *lipid peroxides* (or fatty free radicals). These lipid peroxides are toxic to the cell and highly reactive, leading to free-radical damage of proteins and mtDNA. Here, the positive feedback loop rears its ugly head again. Lipid accumulation is a cause of lipotoxicity and leads to mitochondrial dysfunction through oxidative damage. On the other hand, mitochondrial damage promotes the accumulation of lipids, which cannot be metabolized, and promotes further lipid accumulation.

To prevent the development of this positive feedback loop, a protective mechanism is normally present in healthy cells. *Uncoupling protein 3* (UCP3), located on the mitochondrial membrane, plays a major role, essentially acting as an overflow valve, so that an excessive proton gradient doesn't slow down the ETC. However, research shows dysfunctional UCP3 leads to free-radical damage in cells, which is associated with insulin resistance and type 2 diabetes. This is an active area of diabetes research.

So at this point, the chain of events looks something like this: (1) mitochondrial damage in the target cells, such as muscles, results in lipid accumulation; (2) lipid accumulation results in insulin resistance; (3) due to insulin resistance, the beta cells in the pancreas must increase their metabolism to create more insulin (and then package it up, and secrete it—all of which takes energy); (4) while this metabolism increase helps control blood sugar to some degree in the short term, over time these beta cells accumulate damage to their mitochondria due to chronically high metabolism and energetic demand; and finally, (5) the beta cells start to die off, resulting in a drop in insulin and a spike in blood glucose, which is what is typically seen in long-standing uncontrolled type 2 diabetes.

Mitochondrial Diabetes

Typically presenting at middle age, mitochondrial diabetes is a form that originates from an mtDNA defect, so this diabetes is maternally transmitted and, interestingly, often associated with hearing loss (particularly for high tones). This type of diabetes is characterized by decreased insulin secretion but not insulin resistance, suggesting that the major problem is with the mitochondria of pancreatic beta cells.

The underlying clinical pathology of mitochondrial diabetes might seem to resemble that of type 1 diabetes, but in this case, the immune system doesn't destroy beta cells.

Instead, this type of diabetes results from mtDNA mutations. The most common mutation leading to mitochondrial diabetes is one that encodes for "transfer RNA." The defect in transfer RNA leads to impaired synthesis of multiple mitochondrial proteins and, ultimately, mitochondrial dysfunction. Although rare, its presentation will often resemble that of type 2 diabetes, and it must be diagnosed properly in order to treat it properly. (You can see why this is often misdiagnosed when its pathology resembles that of type 1 diabetes, while its symptoms resemble those of type 2 diabetes.)

Medication-Induced Mitochondrial Damage and Disease

With the increasing use of pharmaceutical medications, there has been a rise in health conditions linked to mitochondrial dysfunction—and mitochondrial dysfunction is increasingly implicated in the etiology of drug-induced toxicities. Despite this, testing for mitochondrial toxicity is still not required by the Food and Drug Administration of the United States, Health Canada, or any other regulatory body responsible for drug approval. Medications can damage the mitochondria both directly and indirectly (see table 2.1

on page 96); they can directly inhibit mtDNA transcription of ETC complexes (the thirteen key subunits), damage ETC components through other mechanisms, or inhibit enzymes required for any of the steps of glycolysis and beta-oxidation. Indirectly, medications could cause mitochondrial damage through the production of free radicals, by decreasing the quantity of endogenous antioxidants such as superoxide dismutase and glutathione, or by depleting the body of nutrients required for the creation or proper function of ETC complexes or mitochondrial enzymes.

In fact, damage to mitochondria might explain the side effects of many medications. Barbiturates (used as sedatives or antianxiety drugs) were the first medications noted to impede mitochondrial function by inhibiting Complex I. This same mechanism also explains how rotenone (an agricultural pesticide) causes mitochondrial damage (which, incidentally, happens to make this a useful chemical for inducing Parkinson's disease in animals so we can study them). Other drugs can sequester coenzyme A (e.g., aspirin, valproic acid), inhibit biosynthesis of CoQ10 (e.g., statins), deplete antioxidant defenses (e.g., acetaminophen), inhibit mitochondrial beta-oxidation enzymes (e.g., tetracyclines, several anti-inflammatory drugs), or inhibit both mitochondrial beta-oxidation and oxidative phosphorylation (e.g., amiodarone). Other substances impair mtDNA transcription or replication. In severe cases, impairment of energy production could contribute to liver failure, coma, and even death.

Many psychotropic medications also damage mitochondrial function. These drugs include antidepressants, antipsychotics, dementia medications, seizure medications, mood stabilizers such as lithium, and Parkinson's disease medications.

Adverse effects of the antiretroviral drugs for treating AIDS result from inhibition of the enzyme responsible for mtDNA replication. Inhibiting this enzyme can lead to a decrease in mtDNA, the thirteen critical subunits of the ETC, and, ultimately, cellular energy production. Mitochondrial dysfunction induced by these drugs explains the many adverse reactions reported for them, including polyneuropathy, myopathy, cardiomyopathy, steatosis, lactic acidosis, pancreatitis, pancytopenia, and proximal renal tubule dysfunction.

TABLE 2.1. Medications Documented to Induce Mitochondrial Damage

Drug Class	Drugs
Alcoholism medications	disulfiram (Antabuse)
Analgesic (for pain) and anti-inflammatory	acetylsalicylic acid (aspirin), acetaminophen (Tylenol), diclofenac (Voltaren, Voltarol, Diclon, Dicloflex Difen, Cataflam), fenoprofen (Nalfon), indomethacin (Indocin, Indocid, Indochron E-Rm, Indocin-SR), naproxen (Aleve, Naprosyn)
Anesthetics	bupivacaine, lidocaine, propofol
Angina medications	perhexiline, amiodarone (Cordarone), diethylami-noethoxyhexestrol (DEAEH)
Antianxiety medications	alprazolam (Xanax), diazepam (Valium, Diastat), barbiturates, amobarbital (Amytal), aprobarbital, butabarbital, butalbital (Fiorinal), hexobarbital (Sombulex), methylphenobarbital (Mebaral), pentobarbital (Nembutal), phenobarbital (Luminal), primidone, propofol, secobarbital (Seconal), talbutal(Lotusate), thiobarbital
Antiarrhythmic	amiodarone (Cordarone)
Antibiotics	tetracycline, antimycin A, fluoroquinolone (Cipro, Factive, Levaquin, Avelox, Noroxin, Floxin)
Antidepressants	amitriptyline (Lentizol), amoxapine (Asendis), citalopram (Cipramil), fluoxetine (Prozac, Symbyax, Sarafem, Fontex, Foxetin, Ladose, Fluctin, Prodep, Fludac, Oxetin, Seronil, Lovan)
Antipsychotics	chlorpromazine, fluphenazine, haloperidol, risperidone, quetiapine, clozapine, olanzapine

Acetaminophen, the popular over-the-counter pain reliever and antifever medication, is the active ingredient in more than one hundred products. This drug is the leading cause of drug-induced liver failure in the United States. Each year more than 450 deaths are caused by acute and chronic acetaminophen toxicity. Acetaminophen is metabolized in the liver, and when acetaminophen passes through

Drug Class	Drugs
Cancer (chemotherapy) medications	mitomycin C, profiromycin, adriamycin (also called doxorubicin and hydroxydaunorubicin and included in the following chemotherapeutic regimens—ABVD, CHOP, and FAC)
Cholesterol medications	**Statins:** atorvastatin (Lipitor, Torvast), fluvastatin (Lescol), lovastatin (Mevacor, Altocor), pitavastatin (Livalo, Pitava), pravastatin (Pravachol, Selektine, Lipostat), rosuvastatin (Crestor), simvastatin (Zocor, Lipex) **Bile acids:** cholestyramine (Questran), clofibrate (Atromid-S), ciprofibrate (Modali), colestipol (Colestid), colesevelam (Welchol)
Dementia medications	Tacrine (Cognex), Galantamine (Reminyl)
Diabetes medications	metformin (Fortamet, Glucophage, Glucophage XR, Riomet), troglitazone, rosiglitazone, buformin
Epilepsy/seizure medications	valproic acid (Depacon, Depakene, Depakene syrup, Depakote, depakote ER, depakote sprinkle, divalproex sodium)
HIV/AIDS medications	atripla, Combivir, Emtriva, Epivir (abacavir sulfate), Epzico, Hivid (ddC, zalcitabine), Retrovir (AZT, ZDV, zidovudine), Trizivir, Truvada, Videx (ddI, didanosine), Videx EC, Viread, Zerit (d4T, stavudine), Ziagen, Racivir
Mood stabilizers	lithium
Parkinson's disease medications	tolcapone (Tasmar), Entacapone (COMTan, also in the combination drug Stalevo)

the enzyme that starts its elimination, it is metabolized to a toxic intermediate that is subsequently neutralized by glutathione before finally being excreted in urine. Therefore, the earliest effect of acetaminophen poisoning is a depletion of the liver's glutathione, the accumulation of free radicals, and decreased mitochondrial function. Because glutathione depletion is a mechanism by which acetaminophen causes death of liver cells, it is not surprising that the antidote for acetaminophen poisoning is a common nutritional supplement called N-acetyl-cysteine (the precursor of glutathione), which increases glutathione.

The exact mechanism of mitochondrial damage and the tissues affected depend on the medication. For example, valproic acid depletes L-carnitine, which causes a decrease in beta-oxidation in the liver, thereby contributing to fatty liver. The antipsychotic medications in table 2.1 inhibit ETC function. The antianxiety medication diazepam was shown to inhibit mitochondrial function in the brain, while alprazolam does so in the liver. Long-term administration of corticosteroids has been shown to result in mitochondrial dysfunction and oxidative damage to mtDNA (and nDNA as well).

If I had my way, all medications would be studied for their effect on mitochondrial function. In fact, all chemicals, whether they're pesticides, food additives, or personal care products (and everything else), should be. For example, the artificial blue color often used in candies and men's shaving gels inhibits oxidative phosphorylation. Table 2.1 lists medications documented to induce mitochondrial damage to date.

Mitochondrial Disease

While I'm embarrassed to admit that I used to be an avid viewer of the television show *The Bachelor*, it was exciting to see that on Season 17, Episode 3 (aired January 2013), Sean (the bachelor) and Ashlee (a "contestant") went on a date with two girls with mitochondrial disease. For many of you, if you watched the show, it might have been the first time you heard of mitochondrial disease. However, this group of illnesses is becoming more and more recognized as genetic testing and genetic sequencing technology become easier, cheaper, and more readily available.

Before the human mitochondrial genome was completely sequenced in the early 1980s, reports of mitochondrial diseases were rare. Since then, it's become possible to sequence the mtDNA of many patients. This has resulted in the number of cases skyrocketing, with about one in five thousand (or even one in every 2,500) people reportedly born with a mitochondrial disease (which might actually be an underestimation, because many are mild forms that don't need medical intervention). Further, the number of conditions recognized as mitochondrial disease has also skyrocketed, and is revealing the bizarre nature of the disease.

Mitochondrial diseases present extraordinarily complicated genetic and clinical pictures that cut across an extremely broad range of established diagnostic categories. The patterns of inheritance do not follow Mendel's laws, although sometimes they do. (Mendel's laws govern the pattern of inheritance of "normal" nuclear genes. The likelihood of a genetic trait or disease being inherited is easily calculated based on the probability of inheriting one of two random copies of the same gene from each parent [giving everyone two copies of each gene].) When mitochondrial diseases are a result of defective nuclear genes, the inheritance patterns do follow Mendel's laws. However, there are two sets of genomes involved in making mitochondria work, mtDNA (inherited only from the mother) and nDNA (inherited from both parents), so inheritance patterns can range greatly from autosomal recessive to autosomal dominant to maternal inheritance.

Making matters more complicated, there are countless interactions between the mtDNA and the nDNA in a cell. The result is that the same mtDNA mutations can produce remarkably different symptoms between siblings in the same family (they will have different nDNA, even though they might all share the same mtDNA), while different mutations can produce the same symptoms. Even twins with the same disease can have a radically diverse symptom picture (the symptoms are related to what tissues are affected), while individuals with different mutations might present with the same symptoms and disease.

However, the variation in mtDNA in the mother's eggs is surprisingly high, and this fact throws a wrench in any predictions of inheritance. To illustrate the bizarre nature of this group of diseases, the onset of symptoms can vary by decades, and even between siblings with the identical genetic mitochondrial mutation. Moreover, occasionally the disease might even disappear in an individual who had (or should have) inherited them. This individual is a lucky one, however, because in general, mitochondrial diseases become progressively worse with

TABLE 2.2. Signs, Symptoms, and Diseases Associated with Mitochondrial Dysfunction

ORGAN SYSTEM	POSSIBLE SYMPTOM OF DISEASE
Muscles	hypotonia, weakness, cramping, muscle pain, ptosis, ophthalmoplegia
Brain	developmental delay, autism, dementia/Alzheimer's disease, seizures, neuropsychiatric disturbances, atypical cerebral palsy, migraines, stroke
Nerves	neuropathic pain/weakness, inflammatory demyelinating polyneuropathy, absent deep tendon reflexes, neuropathic gastrointestinal problems (GERD, constipation, bowel pseudo-obstruction), fainting, absent or excessive sweating, abnormal temperature regulation
Kidneys	Fanconi syndrome, possible loss of protein (amino acids), magnesium, phosphorus, calcium, and other electrolytes
Heart	cardiac conduction defects (heart blocks), cardiomyopathy
Liver	hypoglycemia, gluconeogenic defects, nonalcoholic liver failure
Eyes	optic neuropathy and retinitis pigmentosa
Ears	sensorineural hearing loss, aminoglycoside sensitivity
Pancreas	diabetes and exocrine pancreatic failure
Systemic	failure to thrive, short stature, fatigue, respiratory problems

time. Tables 2.2 and 2.3 outline the symptoms and diseases associated with mitochondrial dysfunction, as well as inherited conditions that implicate it.

Currently, there are over two hundred known types of mitochondrial mutations, and a wide range of common degenerative diseases has been found to involve one or more of these mutations (indicating we might need to reclassify a very large number of diseases as mitochondrial diseases).

Acquired Conditions That Implicate Mitochondrial Dysfunction

As our understanding of mitochondrial function and dysfunction expands, we start to realize the wide range of conditions where mitochondrial dysfunction lies at the heart of many familiar diseases. Some recent data suggests mitochondrial disease is now seen in as many as one in every 2,500 people. When you consider the following conditions, it's realistic to predict that we could see the reported incidence of mitochondrial disease (inherited or acquired) skyrocket to more than one in twenty, or even one in ten.

- Type 2 diabetes
- Cancers
- Alzheimer's disease
- Parkinson's disease
- Bipolar disorder
- Schizophrenia
- Aging and senescence
- Anxiety disorders
- Nonalcoholic steatohepatitis
- Cardiovascular diseases
- Sarcopenia (loss of muscle mass and strength)
- Exercise intolerance
- Fatigue, including chronic fatigue syndrome, fibromyalgia, and myofascial pain

As we know, these mutations cause the mitochondria to fail in their task of making energy. When less and less energy is made in the cells, the cells might stop working or die. All cells (except red blood cells) contain mitochondria, so mitochondrial diseases tend to affect

TABLE 2.3. Inherited Conditions That Implicate
Mitochondrial Dysfunction

Condition	Symptoms
Kearns-Sayre syndrome (KSS)	external ophthalmoplegia, cardiac conduction defects, and sensorineural hearing loss
Leber hereditary optic neuropathy (LHON)	visual loss in young adulthood
Mitochondrial encephalomy-opathy, lactic acidosis, and strokelike syndrome (MELAS)	varying degrees of cognitive impairment and dementia, lactic acidosis, strokes, and transient ischemic attacks
Myoclonic epilepsy with ragged-red fibers (MERRF)	progressive myoclonic epilepsy
Leigh syndrome	subacute sclerosing encephalopathy, seizures, altered states of consciousness, dementia, ventilatory failure
Neuropathy, ataxia, retinitis pigmentosa, and ptosis (NARP)	dementia, in addition to the symptoms described in the acronym
Myoneurogenic gastrointestinal encephalopathy (MNGIE)	gastrointestinal pseudo-obstruction, neuropathy

multiple body systems (either at the same time, or progressively at different times). Of course, some organs and tissues require more energy than others. When an organ's energy requirements can no longer be fully met, the symptoms of mitochondrial disease manifest. They primarily affect the brain, nerves, muscles, heart, kidneys, and endocrine system—all organs with high demand for cellular energy.

At the genetic level, it gets increasingly complex. An individual's *bioenergetic baseline* might be established by understanding the level of inherited defects in mtDNA. As additional mtDNA defects develop over the course of a lifetime, their bioenergetic capacity might decline until an organ's threshold is crossed, where the organ begins to malfunction or become susceptible to degeneration (each organ has a different threshold—which I'll discuss in a moment).

Another genetic complication is that each mitochondrion contains up to ten copies of mtDNA, and each cell and tissue contains many mitochondria, which means there could be countless different defects in the countless different copies of mtDNA in each cell, tissue, or organ. For a particular tissue or organ to become dysfunctional, a critical number of its mtDNAs must be defective. This is called the *threshold effect*. Each organ or tissue is more susceptible to some mutations than others and has its own particular mutational threshold, energy requirement, and sensitivity to free-radical damage. All these factors combine to determine how it will respond to genetic damage.

For example, if only 10 percent of the mitochondria are dysfunctional, the 90 percent that are healthy might mask the effects of their defective counterparts. Or perhaps it's a minor, less serious mutation, but present in a larger percentage of mitochondria. In that case, the cell might still be able to operate normally.

Then there is the concept of segregation. When a cell divides, its mitochondria are distributed at random between the two daughter cells. One might inherit all the defective mitochondria, and the other might get all the healthy ones (or any combination in between). The cell with the defective mitochondria will die by apoptosis, leaving the healthy cell (which could explain how mitochondrial disease can sometimes randomly and unexpectedly disappear). This phenomenon, where not all of the mtDNA copies within a cell are the same, is known as *heteroplasmy* (meaning mixed mtDNA). The degree of heteroplasmy will differ from person to person, even within a family. The degree of heteroplasmy within a person might also differ from organ system to organ system and even cell to cell, leading to a vast array of possible disease presentations and symptoms.

In a developing embryo, as cells divide, different mitochondria with different mutations will eventually populate different tissues with different metabolic requirements. If the defective mitochondria happen to get distributed to cells that eventually develop into metabolically active tissues, such as the heart or brain, these individuals will likely have a poor quality of life, if any. On the flip side, if the defective mitochondria end up in less metabolically active cells, such as skin cells (that are also constantly shed), then the individual might never know they have the genetic imprint for a mitochondrial disease. This was evident on *The Bachelor* when one of the girls affected by mitochondrial disease appeared, on the surface, to be "normal," while the other was affected to a very noticeable degree.

Some specific mitochondrial mutations develop spontaneously with age, due to free-radical production from normal metabolism, which generates a heterogeneous population of mitochondria in the affected cell. What happens next depends on a number of factors. For example, if the affected cell were a rapidly dividing one, such as a stem cell that regenerates tissue, the defective mitochondria would be replicated and its prominence would spread. If the affected cell were one that was no longer active in cell division, such as a nerve cell, then the mutation would be contained to that one cell. I shouldn't rule out the possibility that a random mutation could be beneficial, which does happen occasionally. However, the genetic complexity of mitochondrial disease helps explain how bioenergetic decline, driven by mitochondrial mutations, can have such varied and complex effects over the course of aging.

We must also remember that there are numerous other genes involved in the proper functioning of mitochondria. If a mutation affects a gene coding for RNA, the consequences are usually serious. If a mutation affects a mitochondrial transcription factor in the nDNA inherited from either parent at conception, then the effects could affect all the mitochondria in the body. Yet, if the mutation affects certain mitochondrial transcription factors that are only activated in a particular tissue, or in response to a particular hormone, the effects would be tissue-specific.

Given this wide degree of variance, the incredibly broad spectrum of mitochondrial diseases is a problem in and of itself and makes it almost impossible to predict how the disease will progress in any one individual. Also, there are so many types of mitochondrial disease it would be difficult to name them all, and many have yet to be discovered. Even many common degenerative diseases (e.g., various cardiovascular diseases, cancers, various forms of dementia) are now being seen as specific mitochondrial diseases.

While there is no cure for genetic mitochondrial disease, many people—especially those with milder forms—might have a normal quality of life and life span if their disease can be well managed.

When Mitochondrial Disease Is the Primary Disease

When mitochondrial disease exists right from birth, it is said to be the *primary disease.* In mild cases, young people might learn to cope and adapt to the amount of energy they have and don't realize they even have mitochondrial disease. Mildly affected adults might say that they had a very normal and healthy life as a child, although they were never really good at sports or activities that required endurance.

Some common symptoms of more noticeable mitochondrial disease include developmental delay or regression in development, seizures, migraine headaches, muscle weakness (might be only occasional), poor muscle tone (*hypotonia*), poor balance (*ataxia*), painful muscle cramps, inability to keep up with peers in physical activities (low endurance), chronic fatigue, stomach problems (vomiting, constipation, pain), temperature problems from too little or too much sweating, breathing problems, eyes that are not straight (*strabismus*), decreased eye movement (*ophthalmoplegia*), loss of vision or blindness, droopy eyelids (*ptosis*), loss of hearing or deafness, heart/liver/kidney disease at a young age, and shaky parts of the body (tremors). Although some of these symptoms are common in the general population, people with mitochondrial diseases are usually affected with multiple symptoms at a young age.

Similarly, many common health conditions can be misdiagnosed. While the diagnosis could be relevant to the symptomatology, the underlying condition might actually be mitochondrial disease. For example, mitochondrial defects could contribute to a diagnosis of heart disease in some patients. A recent study of dilated cardiomyopathy found that about one in four patients (25 percent) had mtDNA mutations in the heart tissue.

Other patients might have a genetic CoQ10 deficiency and suffer dysfunctions in the brain, nerves, and muscles, often including fatigue

TABLE 2.4. Features of Mitochondrial Disorders Caused by mtDNA Mutations (Maternally Inherited)

CHILDREN	
Cardiac	biventricular hypertrophic cardiomyopathy, rhythm abnormalities, cardiac murmur, sudden death
Skin	erythema (redness), lipomatosis (benign fatty deposits), reticular pigmentation ("dark dot" disease), hypertrichosis (excessive hair growth), vitiligo (patchy depigmentation), alopecia (hair loss)
Endocrine	diabetes mellitus, adrenal failure, growth failure, hypothyroidism, hypogonadism (sex organs not producing enough hormones), hypoparathyroidism
Gastrointestinal	failure to thrive, dysphagia (difficulty swallowing), GI motility problems, vomiting, pseudo-obstruction
Blood	anemia, pancytopenia (low red and white blood cells)
Hepatic	hepatic (liver) failure
Musculoskeletal	weakness, myopathy (muscle weakness)
Neurological	myopathy (muscle weakness from dysfunctional nerves), developmental delay, ataxia (uncoordinated muscle movements), spasticity, dystonia (involuntary muscle contractions), hypotonia (low muscle tone), bulbar signs, chorea (an abnormal movement disorder), seizures, myoclonus (involuntary twitching), stroke
Eyes	optic atrophy, retinitis pigmentosa (damaged retina), ptosis (droopy eyelids), diplopia (double vision), cataract
Ears	sensorineural deafness
Kidneys	renal tubular defects, nephrotic syndrome, tubulointerstitial nephritis
Respiratory	central hypoventilation, apnea (temporary cessation of breathing)

ADULTS	
Cardiac	heart failure, conduction block, cardiomyopathy, sudden death
Endocrine	diabetes, thyroid disease, parathyroid disease
Gastrointestinal	constipation, irritable bowel syndrome, dysphagia, anorexia, abdominal pain, diarrhea
Musculoskeletal	rhabdomyolysis (breakdown of muscle fibers), muscle weakness, exercise intolerance
Neurological	migraine, stroke, seizures, dementia, myopathy, peripheral neuropathy, ataxia, speech disturbances, bulbar signs, myoclonus, tremor
Eyes	optic atrophy, cataract, progressive external ophthalmoplegia, ptosis, pigmentary retinopathy, vision loss, diplopia
Ears	sensorineural deafness
Reproductive	pregnancy loss in mid to late gestation, hypogonadism
Respiratory	respiratory failure, nocturnal hypoventilation, recurrent aspiration, pneumonia

on exertion, and seizures. Such patients appear to respond to CoQ10 supplementation, but observations are limited because diagnosis of this disorder is in its infancy. Primary CoQ10 deficiency is one of the mitochondrial diseases caused by mutations in nDNA.

Table 2.4 outlines the features of mitochondrial disorders caused by mtDNA mutations in children and adults.

Treating Mitochondrial Disease

Recent advances in medicine, science, and genetics have helped us gain a better understanding of the diagnosis and treatment of mitochondrial diseases. Unfortunately, there is still no cure, and current treatments cannot guarantee symptom relief or improvement in quality of life.

The effectiveness of treatment varies depending on the exact disorder affecting a person and its severity. Also, treatment cannot reverse any existing damage (such as brain malformations or damage from a stroke). Generally, those with a mild form of the disease tend to respond to treatment better than those with a severe form.

Recent research has shown that several nutritional supplements can help relieve symptoms and improve function. Case reports and pilot studies have found that some patients with mitochondrial diseases respond to long-term CoQ10 therapy, and promising results have been reported in mitochondrial encephalomyopathy, lactic acidosis, and strokelike syndrome (MELAS); Kearns-Sayre syndrome;

Examples of Inherited Disorders Caused by mtDNA Mutations

The following disorders have been shown to be caused by mutations in mtDNA, which means they will typically follow maternal inheritance patterns.

- Mitochondrial encephalomyopathy, lactic acidosis, and strokelike syndrome (MELAS)
- Myoclonic epilepsy associated with ragged-red fibers (MERRF)
- Neuropathy, ataxia, and retinitis pigmentosa
- Maternally inherited Leigh syndrome
- Leber hereditary optic neuropathy (LHON)
- Chronic progressive external ophthalmoplegia
- Maternally inherited diabetes and deafness
- Nonsyndromic maternally inherited deafness
- Kearns-Sayre syndrome (KSS)
- Pearson syndrome

and maternally inherited diabetes with deafness. An Italian study of patients with mitochondrial disease measured the bioenergetic activity in their brain and skeletal muscles, and demonstrated the positive impact of CoQ10 therapy. After six months of CoQ10 therapy (at only 150 milligrams daily), brain bioenergetics returned to normal in all patients, and skeletal muscle energetics improved significantly. A number of other studies have also confirmed the value of CoQ10 in mitochondrial disease. Typically, it's prescribed in combination with other nutrients, in what's often called the "mitochondrial cocktail," and a knowledgeable and progressive doctor might recommend some or all of these supplements: creatine monohydrate, vitamin C, vitamin E, alpha-lipoic acid, thiamine (vitamin B_1), riboflavin (B_2), niacin (B_3), L-carnitine, or L-arginine. Others include D-ribose, PQQ, magnesium, and medium-chain fatty acids.

Regular exercise and physical activity provide immense benefits for the body, mind, and spirit for everyone. However, for those with mitochondrial disease, physical activity becomes extremely important as a way to build more mitochondria in the cells (mitochondrial

Examples of Mitochondrial Disorders Caused by nDNA Mutations

These disorders are known to be caused by mutations in nDNA, which means they can be inherited either maternally or paternally, and will typically follow Mendel's inheritance pattern.

- Autosomal recessive external ophthalmoplegia (paralysis of muscles that control eye movements)
- Hypertrophic cardiomyopathy (large, damaged heart)
- Myoneurogastrointestinal encephalomyopathy
- Leigh syndrome
- Mitochondrial depletion syndrome
- Dominant optic atrophy

biogenesis), which generates more energy for the cells, resulting in increased tolerance to physical activity. Ultimately, exercise can improve the quality of life for anyone with a mitochondrial disease.

Age-Related Hearing Loss

Age-related hearing loss (known as *presbyacusia* or *presbycusis*) affects approximately one-third of all people aged sixty-five and older, and is due to the changes that occur in the aging body. Circulatory disorders, for example, which limit the flow of blood throughout the body, as well as to the brain and auditory system, are common in later years. There are many reasons why circulation slows down as the body grows older, among them heart disease, hardening of the arteries, diabetes, and sedentary lifestyles.

Alarmingly, nearly half of all baby boomers today suffer from some degree of hearing loss. While its onset can be almost imperceptible, the end result is an impaired ability to interact with the world that significantly detracts from quality of life. In fact, "cognitive load" is a possible explanation from newer research that shows that hearing loss is linked to cognitive decline. It's possible that as hearing becomes worse, a person's perceived intelligence decreases because, during a conversation, the brain spends more resources just trying to hear what's being said, rather than processing the contents of what's been spoken.

But how do mitochondria fit into this explanation? A growing body of evidence from animal studies has suggested that the cumulative effect of free radicals could induce damage to mtDNA, which can result in the eventual apoptosis of the cochlear cells of the ear. Exposure to noise is known to induce excess generation of free radicals in the cochlea, and genetic investigation has identified several genes that mutate, including those related to antioxidant defense and atherosclerosis. Genetic variation in the antioxidant defense system might also help explain the large variations in the onset and extent of age-related hearing loss among elderly subjects.

What's fascinating is that the research and clinical work of a leading otolaryngologist show that it is possible to slow the progression of—and sometimes even reverse—hearing loss using an integrative approach that includes optimal nutritional and lifestyle choices. In his book, *Save Your Hearing Now*, Michael Seidman revealed that age-related hearing loss is linked to free-radical damage and mitochondrial dysfunction.

After establishing this link, Seidman tested the theory using caloric restriction in rats. A calorie-restricted diet (which I will detail later in chapter 3, "Ketogenic Diets and Calorie Restriction," page 170) is proven to reduce free-radical production and reduce mitochondrial damage. Compared to rats that were allowed to eat freely, the calorie-restricted rats and another group treated with antioxidants (including melatonin and vitamins E and C) had decreased progression of age-related hearing loss.

Another study showed that elderly rats given either acetyl-L-carnitine or alpha-lipoic acid saw an improvement in hearing, compared to the placebo group (which saw further decrease in hearing over the study period). Our understanding of how these nutrients function in relation to mitochondria indirectly confirms the organelle's involvement in age-related hearing loss. Along with improved hearing in the supplemented rats, the researchers found a much lower level of mitochondrial damage all throughout the body. The supplements actually reduced the amount of free-radical damage everywhere, creating an antiaging effect that not only improved hearing but also carried over to other cells throughout the body.

A study published in 2013 looked at a Hungarian family with numerous members suffering from unexplained hearing loss. Based on the matrilineal inheritance of this condition, the researchers analyzed the family's mtDNA to discover what was leading to this hearing impairment. It was a genetic mutation within mtDNA, which adds significant weight to the theories about mitochondria's role in at least a portion of those with hearing loss.

Mitochondria, Aging Skin, and Wrinkles

It might seem politically incorrect or superficial to say it, but your face is one of your most precious assets. Unfortunately, as we age, most people will display a face ravaged by time, free-radical damage, and excessive sun exposure. And while some might *feel* young and vibrant, their face might communicate a very different state of affairs.

The skin is not only the body's largest organ, it is also one of the most complex, comprising multiple layers of epithelial tissues that protect underlying muscles and organs. The skin also has a multitude of functions: It protects against pathogens; provides insulation, temperature regulation, and sensation; and helps synthesize vitamin D. The *epidermis* (the outermost layer of skin) provides a waterproof barrier against the external environment, and largely contains *keratin* (a fibrous protein made by keratinocyte cells) and *melanin* (the skin's main pigment, which is produced by melanocyte cells).

Below the epidermis is the *dermis*. In addition to providing essential support to the epidermis, the dermis contains nerves, glands, and essential proteins called *collagen* and *elastin*. Collagen is the skin's main structural protein, while elastin provides elasticity to the skin. This layer also contains essential fats and *glycosaminoglycans*, which are large, sugarlike molecules that bind and hold water—all helping to keep moisture in the skin.

Emerging research now suggests that impaired mitochondrial energy production plays an important role in aging of the skin. For example, in aging adults, fibroblast cells demonstrate dramatic mitochondrial dysfunction. Fibroblasts are cells essential to youthful, healthy skin; they produce collagen and elastin.

With mitochondrial dysfunction, fibroblasts are less capable of producing the energy required to carry out their essential skin-related functions of manufacturing collagen and elastin. Scientists believe that this energy deficit in fibroblast cells contributes to the visible signs of skin aging, which might be why so many antiaging creams and concoctions now contain CoQ10, an essential component of the ETC.

Further, years of cumulative free-radical damage (from whatever source, including ultraviolet radiation from excessive sun exposure) can induce mitochondrial dysfunction. This dysfunction eventually leads to dramatic changes in the skin's health and appearance.

The epidermis becomes less capable of tissue repair and renewal. Collagen becomes sparser and less soluble. Elastin fibers are slowly degraded and damaged, and areas of sun-damaged skin accrue abnormally structured elastin. Glycosaminoglycans can no longer properly interact with water, while lipid content decreases with age. The end result of these age-related changes is that skin becomes prone to wrinkling, dryness, sagging, decreased flexibility, dullness, and poor healing responses.

Infertility and Mitochondria

As discussed, in general, for most organisms, including mammals, mtDNA from the father is not transmitted to the offspring. Only maternal mitochondria, already present in the oocyte, remain. Therefore, the study of mitochondria in women has provided incredible insight into mitochondrial dysfunction and its consequences.

It's well established that as women age, the risk of infertility increases. Further, should a woman get pregnant at an older age, there is a much higher risk of birth defects. Unfortunately, "older" with respect to fertility means over thirty-five years of age (although still "young" by any other account).

The reason why mitochondria are so important in fertility is that each oocyte contains about one hundred thousand mitochondria—a huge number despite the lack of demand (most ooctyes are dormant for the majority of their existence, presumably to protect mitochondria from damage for as long as possible). In contrast, a sperm cell only has a few hundred mitochondria.

To be motile, sperm need to have very high metabolisms. Unfortunately, this results in significant free-radical damage in their very short life spans, and sperm have been known to quickly accumulate mutations in their mtDNA. By eliminating this potential source of defective DNA, the oocyte prevents such mutations from being passed on and affecting the offspring.

Within minutes after fertilization, if and when sperm-derived mitochondria enter the oocyte, a localized reaction is triggered around the sperm's mitochondria, where the autophagosomes engulf the paternal mitochondria, resulting in their degradation (referred to as *autophagy*). This ensures only maternal inheritance of mtDNA. However, in a situation where autophagy is impaired, paternal mitochondria and their genome remain even in the first stages of embryonic development.

Another reason why this mixing of mtDNA is so disastrous and why the oocyte eliminates the paternal mtDNA lies in the way proteins responsible for energy production are created. Remember that the mtDNA encode for only thirteen proteins related to the ETC. By comparison, the DNA in the nucleus codes for over eight hundred proteins involved in the respiratory chain. Further, even if the mtDNA is "normal," the newly created nDNA (which never existed before, and is a mix of both parents' nDNA) must not only be "normal" itself, but must also communicate efficiently with its mtDNA counterpart.

For example, for Complex I to be built properly, there needs to be clear communication between the mtDNA and nDNA. However, it's the mtDNA that's in charge because it codes for the critical subunits of the complexes, which get embedded into the inner mitochondrial membrane. Once there, it acts as a beacon that attracts the other subunits encoded by nDNA. If the mtDNA and nDNA are a good "fit," the complexes get built properly and in sufficient quantities. This allows the production of sufficient energy that's "clean" (meaning, there's minimal free-radical leakage), and the mitochondria survive. As long as there are enough of these types of mitochondria in a cell, the cell survives.

On the flip side, if they are not a good fit and energy production is suboptimal, the mitochondria die. If there are enough of these types of mitochondria in a cell, the cell dies. For this reason, ensuring only one set of mtDNA survives in the fertilized egg makes this whole process smooth.

Oocytes from women of advanced reproductive age show accumulations of mtDNA mutations that impair function. Vast amounts of energy are needed for the rapid cell division in the embryo, so mtDNA mutations aren't compatible with the prospects of normal and healthy fertility, and any fertilization of these eggs will quickly be terminated (i.e., result in a miscarriage).

Based on this, infertile women in their thirties (or even forties) who were otherwise healthy could be candidates for a procedure called *nuclear genome transfer*. Remember, this is where an oocyte from a healthy and fertile female donor has its nucleus removed (leaving all other components, including the healthy mitochondria), and then the nucleus from the fertilized egg of the infertile woman is transferred into the healthy donor egg.

As mentioned previously, due to ethical issues regarding children born with three parents, this procedure is banned throughout most of the world. Although recent evidence suggests that mitochondrial disease can still creep back in, even if a mother's defective mitochondria are virtually eliminated, one benefit that arose from this experiment in human reproduction was that it essentially confirmed the role of age-related mitochondrial dysfunction as a key contributor to infertility.

The importance of this confirmation is that scientists now have a target to focus on—the mitochondria—and can begin the search for suitable pharmacological compounds that could improve cellular bioenergetics: something that could revive the youthful integrity of healthy mitochondria without using donor mitochondria.

The Purge: Selecting Only the Highest-Quality Eggs and Mitochondria

After the sperm fertilizes the oocyte, the resulting "zygote" goes through tremendous growth through cell division. As previously mentioned, this requires incredible amounts of energy. However, while the cells divide, the mitochondria do not. Instead, the initial number of mitochondria (about one hundred thousand) gets partitioned with each division, so that by a couple weeks after conception, each cell only has about two hundred mitochondria. Again, this is by design. Whereas defective mitochondria could hide and coast by in a sea of healthy mitochondria, when their numbers are reduced to about two hundred per cell, each mitochondrion had better be pulling its own weight. Defective ones can no longer coast alongside their harder-working comrades. When exposed, these dysfunctional mitochondria will be eliminated, which is probably not a big deal if this results in only one cell dying, but when enough mitochondria are defective in enough cells, the pregnancy is terminated.

After all the defective mitochondria and cells have been eliminated (and if that hasn't resulted in a miscarriage), the number of mitochondria per cell can multiply in normal fashion as the embryo grows.

If the embryo is female, she will start to produce her own oocytes at an alarming rate. In fact, at five months' gestation, she's already produced about seven million oocytes! From this crest, the body starts its purging. By the time she's born, she's already down to about two million oocytes. Why? Natural selection—the natural process where only those most able to flourish will survive. During the development of the fetus, the body is comparing its mtDNA to the new nDNA to ensure that any incompatible oocyte is eliminated. It doesn't stop there, though—the purging continues. By puberty, when she is biologically mature enough to bear her own children, she's down to about three hundred thousand oocytes. By this point, only the cream of the crop (in terms of healthy oocytes) remains. Her body, from gestation to puberty, has selected only the healthiest oocytes that will give her the best chance for conceiving her own healthy offspring. From there, the cycle of life repeats itself.

Fertile Coenzyme Q10 (CoQ10)

Biologically, we are most fertile in our adolescence and early twenties. After these optimal reproductive years, there are other things that happen to our bodies that prepare us for our eventual exit from this world—to make room for the next generation and free up valuable resources. One of the things that happens is our body starts to produce less and less CoQ10. Remember, CoQ10 is the compound that shuttles electrons from Complex I (or II) to Complex III in the ETC. As we produce less of this essential compound, we produce less and less cellular energy, starting the chain of events that will ultimately lead to our demise.

However, before our eventual demise, we'll experience various symptoms, and for women, one of them is infertility. If there is a relative deficiency of CoQ10, the eggs cannot produce enough energy, and as a result, when they are fertilized by sperm, the *zygote* (embryo) is aborted.

In cases where there might be sufficient cellular energy production to stay above the threshold for miscarriage, the embryo will continue to develop and mature. Yet, in some cases that evade miscarriage, there is still insufficient cellular energy to properly separate the chromosomes during cell division. An example of this is *Trisomy 21* (more commonly known as Down syndrome), where there are three copies of chromosome 21. This example demonstrates why older women have a higher risk of bearing children with birth defects.

CoQ10 deficiency—either alone or in combination with mtDNA mutations or a mismatch between mtDNA and nDNA—is thought to be responsible for a significant percentage of age-related infertility cases in women. Animal studies have shown promising results, and based on this, it is recommended that women with age-related infertility undergoing various fertility treatments supplement with CoQ10—even in the absence of human studies. As I write this book, however, I'm aware of at least one human clinical trial underway in Toronto, Canada.

Eye-Related Diseases

As we age, we're at a higher risk of developing age-related eye diseases, and we might predict that these conditions have mitochondrial components to them. These conditions include age-related macular degeneration, cataract, glaucoma, diabetic eye disease, and others. I'll discuss a couple of them here.

Age-Related Macular Degeneration

Free-radical damage seems to be the most important factor in the pathogenesis of age-related macular degeneration (AMD), a condition affecting many elderly people in the developed world, and the leading cause of blindness.

Several studies have found an increase in mtDNA damage and a concurrent decrease in the efficacy of DNA repair—and both these factors are correlated with the occurrence and the stage of AMD. Considering that mtDNA repair is executed by proteins encoded by nDNA, certain mutations in the nucleus might affect the ability to repair mtDNA.

It's been suggested that this poor DNA repair, combined with the enhanced sensitivity of retinal cells to environmental stress factors (e.g. ultraviolet light, light in the blue spectrum, air pollution), contributes to the development of AMD. Collectively, the scientific data suggests that the cellular response to both mitochondrial and nuclear DNA damage might play an important role in AMD pathogenesis.

Further, the retina requires proportionally more energy (per cell) than any other tissue in the body, and the retina is also where the density and number of mitochondria per cell are among the highest found in the body. Both of these facts have significant implications for AMD.

With age, the high energy demand and constant assault from ultraviolet and blue light produce debris and significant free-radical damage. In humans, this can result in a decline of up to 30 percent in the numbers of light-receptive cells in the eye by the time we are seventy and can lead to progressively poorer vision.

Glaucoma

Glaucoma is the second-leading cause of blindness. It often progresses without symptoms until it succeeds in damaging the optic nerve. Up to 50 percent of people with glaucoma remain undiagnosed, and as many as one in ten individuals aged eighty or older are afflicted by it. Mitochondrial-associated free-radical damage affects the eyes' drainage system, where tissue integrity is essential to maintaining normal pressure and fluid flow out of the eye.

Encouraging research has shown that several of glaucoma's underlying causative factors might be prevented and even reversed through natural interventions, offering new hope to the millions at risk for this widespread, debilitating condition. The picture is even more encouraging with studies that suggest a common link between glaucoma and Alzheimer's disease. Why? Because, as we discussed earlier, Alzheimer's disease has known deficits in mitochondrial function, and targeting bioenergetics in this cognitive disorder has shown improvements in symptoms and disease progression. So if we can do that for Alzheimer's disease, we should be able to do the same with glaucoma.

Stem Cells Require Healthy Mitochondria

Our bodies possess remarkable abilities for sustained tissue renewal throughout our lifetimes. This continuous self-renewal process is dependent on reservoirs of stem cells.

A report published a few years ago describes how mitochondrial function is crucial for maintenance of stem cells. As age-related mitochondrial dysfunction becomes more prominent, damaging free radicals increase and are accompanied by stem-cell compromise.

Interestingly, researchers have discovered that stem-cell populations do not necessarily decline with advancing age, but instead they lose their restorative potential. This functional stem-cell decline is accompanied by eventual organ malfunction and increased incidence of disease, which bears some resemblance to oocytes and female infertility (where the number of oocytes doesn't fall off with advancing age, but instead they lose their reproductive potential due to declining cellular energy production).

A study published in 2015 by Katajisto and colleagues proved this by observing how mitochondria are portioned off during the formulation of daughter cells. The researchers followed the fates of old and young organelles during division of stemlike cells and saw that aged mitochondria were distributed asymmetrically between daughter cells. Daughter cells that received fewer old mitochondria maintained stem-cell traits, or *stemness*. Further, inhibition of mitochondrial fission disrupted various processes and caused loss of stem-cell properties in the daughter cells. From this study, it appears there might be built-in mechanisms for stemlike cells to asymmetrically sort aged and young mitochondria, allowing at least one daughter cell to maintain a large pool of healthy mitochondria and, as a result, to keep its stemness going as long as possible.

Mitochondrial dysfunction thus underlies a degenerative cycle that robs aging humans of the renewal benefits of their own stem cells. The integration of mitochondria into stem-cell research has led scientists to propose that improvements in mitochondrial health (along with other cellular modulations) could yield advanced therapeutic strategies designed to rejuvenate tissues of the aged.

Cancers: Understanding the Causes Brings Us One Step Closer to Cures

No discussion of mitochondria is complete without a discussion of cancers. As Nick Lane said in *Power, Sex, Suicide*, a cancer is the most extreme example of conflict within an individual—a single cell becomes "selfish," escapes the body's systematic control, and multiplies like a bacterium.

Cancers are usually the result of genetic mutations (although there are exceptions). Typically, a cell must accumulate eight to ten mutations in specific genes before it can transform into a malignant cell—a single mutation is rarely enough. Upon transformation to a malignant cell, its own interests are put before those of its community, the body. There are also several checkpoints in the cells, which is why it takes an average of eight to ten particular mutations before a cell becomes malignant. Some people might inherit some of these mutations from their parents, leaving them with a lower threshold of "new" mutations that must accumulate before the onset of cancer—they have a genetic predisposition to cancer.

Normally, defective cells no longer working toward the greater good are eliminated via apoptosis. Apoptosis is critical to immune function, and helps enable the immune system to distinguish between "self" and "nonself" (where cells that could react against our own body tissues commit apoptosis). Immune system cells can also exert many of their own effects by inducing apoptosis on damaged or infected cells. This kind of active screening by immune cells removes damaged cells before they get a chance to get a foothold and proliferate.

In apoptosis, the chain of events is perfectly coordinated and leaves no evidence that the cell ever existed. However, there is a price to pay for such a coordinated series of events. All steps along the way require ATP—if the supply of ATP fails to meet the cell's demand, the cell cannot commit apoptosis, and the defective cell is given a chance to run wild. Deficient ATP production can happen for many reasons, but two of them are due to gene mutations in either the mtDNA or nDNA that (1) no longer produce functional proteins involved in energy production, or (2) result in defects in any of the many, many proteins involved in the apoptosis cascade. Thus, mitochondrial function is in absolute control over apoptosis and prevention of cancers and, in addition, illustrates how nDNA mutations are in play because it encodes many mitochondrial proteins.

Many people just learning about the importance of mitochondria to human health and aging are shocked to find mitochondria's critical part in cancer formation. What's even more shocking is that this association between mitochondrial dysfunction and cancers was made as early as 1930, when German scientist Otto Warburg first hypothesized that the increased rates of aerobic glycolysis, which he observed in a variety of tumor cell types, might be due to an impaired respiratory capacity in these cells (he was later awarded the Nobel Prize for his research). This increase in aerobic glycolysis changes the bioenergetics of the cell, and the shift has been confirmed by studies showing reduced activity of the TCA cycle and oxidative phosphorylation, increased gluconeogenesis, increased lactic acid production, and reduced fatty acid oxidation. There are some excellent references that summarize the totality of evidence and these are concluding what Warburg knew nearly a century ago: Cancers are a metabolic disorder —just like everything else we're talking about in this book. If this is an area of interest for further reading, I suggest the book *Tripping Over the Truth*, by Travis Christofferson.

Mitochondria as Clinical Markers for Cancers

The abundance of mitochondria and their DNA makes them an attractive molecular marker of cancers. Cancer cells have been found to have 220 more mutated mtDNA than mutated nDNA. Mutant mtDNA is readily detectable in urine, blood, and saliva samples from patients with various cancers, and has been used as a marker in hepatocellular carcinoma and breast cancer. Recently, rapid sequencing protocols have been developed to detect mtDNA variants in tumor and blood samples taken from patients.

Over one thousand different proteins are found in mitochondria, and advances in proteomic technologies have made the quantitative analysis of mitochondrial protein expression possible. In fact, the US National Institutes of Standards and Technology has recently established a mitochondrial proteomic database, and research into mitochondrial protein profiles in normal cells versus cancer cells will hopefully lead to identification of markers for clinical detection of cancers, and contribute to an understanding of how differential protein expression might influence the development of these diseases.

Mitochondria as a Therapeutic Target

There are many distinct differences between the normal structure and function of mtDNA and those of cancer cells, and this fact opens the possibility of targeting cancer cells with anticancer agents. One chemotherapeutic approach utilizes delocalized lipophilic cations (DLCs) that selectively accumulate in cancer cells in response to the increased mitochondrial membrane potential. Several of these compounds have exhibited at least some degree of efficacy in killing cancer cells in the laboratory and in biological systems. Certain DLCs have been employed in *photochemotherapy* (PCT), an investigational cancer treatment involving light activation of a photoreactive drug (also known as a photosensitizer) that is selectively taken up or retained by cancer cells. There has been considerable interest in PCT as a form of treatment for cancers of the skin, lung, breast, bladder, brain, or any other tissue accessible to light transmitted either through the body surface or internally via fiber-optic endoscopes. *Cationic* (or positively charged) photosensitizers are particularly promising as potential PCT agents. Like other DLCs, these compounds are concentrated into the mitochondria of cancerous cells in response to the negative charge inside the matrix. In response to localized photoradiation, the photosensitizer can be converted to a more reactive and highly toxic species, thus enhancing the selective toxicity to carcinoma cells and providing a means of highly specific tumor cell killing without injury to normal cells.

An alternative strategy employs the mitochondrial protein-import machinery to deliver macromolecules to mitochondria. For example, certain short peptides with functional domains act as a sort of homing device that, when internalized into the target cells, readily penetrates the mitochondrial membrane, and becomes toxic by disrupting the proton gradient.

Researchers are also attempting to develop drug and DNA delivery systems attracted to mitochondria. Recent data demonstrates that a *liposome* (a tiny vesicle composed of a phospholipid membrane surrounding another molecule, in this case a chemotherapeutic drug) can be rendered mitochondria-specific via the attachment of known residues to its surface. The goal of this type of research is to someday create mitochondria-specific vehicles that will effectively deliver drugs into the organelle to destroy dysfunctional mitochondria or to replenish mitochondria with healthy copies of the genome (another shot at the Fountain of Youth?).

Aging as a Disease

I conclude this section with aging. Yes, I've talked about this previously, but I want to include it here with all the other "diseases." Why? Because if mitochondrial dysfunction is at the heart of all these age-related degenerative diseases, and I've previously established that deteriorating mitochondrial function is at the fulcrum of the aging process, does it not make sense that aging itself is a disease (just as I wrote in my medical school essay)? What makes this train of thought even more exciting is that if it's a disease, we should be able to "cure" it. Right? In fact, aging can be considered the ultimate disease, because it claims 100 percent of the population (if we don't die from other causes first).

Unfortunately, the mechanisms related to all these degenerative diseases and aging are complex, and numerous. For example, there can be mutations in the protein subunits of the complexes (either in the nDNA or mtDNA), supercomplex disassociation, defective autophagy, defective fission or fusion, defective transcription, defective protein transport or channels, defective protein assembly or folding, defective permeability transition pores, defective caspase or apoptotic enzymes, mutations in any of the enzymes of the TCA cycle or beta-oxidation, mutations in UCP, defects in the peroxisomes or endoplasmic reticulum (or communication errors between these organelles and the mitochondria), inefficient mitochondrial movement, and the list goes on.

So while it seems we can fantasize about the Fountain of Youth, we are still a ways from finding it. The good news is that in the last decade we've made incredible strides, and we don't need to solve the puzzle in one go. To start, all we need to do is extend life span and delay aging. Instead of discovering the Fountain of Youth right off the bat, we just need to make incremental steps toward it. Why? The rate of scientific progress is accelerating—this increases exponentially as technology gives birth to greater technology—and as we are able to slowly extend life span, to say 150 or 160 years, by the time we actually reach that age, science and medical research will have figured out how to get us through the next several centuries.

Research in life extension received a significant boost in September 2013, when Google Inc. announced it was forming and funding a company dedicated to researching aging and age-related diseases. While many people reading this book might say this aspect—life extension—is pretty much impossible, if any organization is going to push the boundaries of what is or isn't possible (at least within our lifetime), Google, with its seemingly bottomless pit of funds and its creative and scientific geniuses, is humankind's best shot.

Google refers to these projects as "moonshots." While they are not part of its main business, moonshots embody the company's philosophy of questioning what's possible, and "10x thinking" (creating things that are at least an order of magnitude better than what already exists). So for Google, these side projects indicate that nothing is off limits—not even death is safe.

There is, of course, debate over how far we would want to extend life. Just extending old age would not be much of an achievement, and it would place a heavy burden on society. The goal should rather be to extend the healthy and vital years of life—or, as a quote I once read said, "the goal is to die young as late as possible." And if we could manage that, there would still be practical questions. For example, the world is already struggling with overpopulation and aging societies.

Extending life could exacerbate the situation. To keep population growth at net-zero, previous generations had to have families with many children. As mortality rates dropped with improved sanitation and health care, we have been able to maintain a slow growth of the population with families having fewer and fewer children. However, even in developed countries, we still find countless families with more than two children. So you can see it would take a significant shift in thinking to keep population growth under control—and it might even require government intervention (such as we saw in China—and you can see how well that worked: badly).

However, before we get into the numerous ethical debates on this topic, let's move on. It is fascinating to discuss, if anything, just from a philosophical point of view, but you should know that there are very good and meaningful debates already in progress among the life-extension community. While "knowledge is power," there is hope in the unknown, and there is just so much we still don't know.

CHAPTER THREE

Nurturing the Force

Nutritional and Lifestyle Factors to Improve Mitochondrial Health

What can we do to preserve our mitochondria and cellular bioenergetics? While most mammals rely heavily on antioxidants to neutralize the free radicals produced within their mitochondria, as Lane discussed in his book, birds slow down the rate of free-radical leakage. Understanding the differences between birds and mammals might help us gain insight into the best way to approach aging and its related degenerative diseases. So let's see how, and if, we can be more like the birds.

How Do the Birds Do It?

It's now confirmed that the majority of free radicals are produced from Complex I of the ETC. The various subunits from which the free radicals leak are positioned in such a way that they leak directly into the matrix, which is where the mtDNA resides. Just about every antioxidant supplement has failed to deliver on its promises because it needs to be targeted to this specific complex with astonishing precision. The next best target would be the matrix to protect the mtDNA.

However, the antioxidants would alter the cells' free-radical signaling. Birds have evolved to approach it from a different, more efficient angle. We already know they have low levels of antioxidants, but how do they reduce free-radical leakage in the first place?

We don't know the definitive answer at this point, but birds might lower their free-radical leakage by uncoupling their ETCs. As I previously mentioned, uncoupling is where electron flow is disconnected from ATP production and, instead, the proton gradient that is created by the electron flow is dissipated as heat. The benefit of uncoupling is that the electrons can continue flowing—they don't get backed up—and this flow in turn reduces free-radical leakage. Uncoupling the proton gradient, in theory, has profound benefits for slowing the progression of all age-related degenerative diseases and also aging itself. It could also help us to burn more calories and lose weight. The promise is promising indeed!

Salicylic acid (or its derivatives, such as aspirin) is a mitochondrial uncoupler, and it has been shown to reduce the risk of a number of degenerative diseases, and even cancers. Because it is a mitochondrial uncoupler, it is also often a part of "stack" formulas frequently used in the fitness and weight-loss industries. However, there are downsides to chronic administration of even low-dose salicylic acid—gastrointestinal ulcers are among the more commonly reported adverse effects. Other uncouplers of note are the recreational drug MDMA (more commonly known as Molly, ecstasy, or "e"), notoriously known for inducing excessive heat production in its users, and metformin, the popular antidiabetes drug, which is now being investigated against a host of other diseases. A thought-provoking study published in June of 2014 demonstrated that metformin consistently benefited African Americans more than their Caucasian counterparts with respect to blood sugar control. Based on our earlier discussion regarding "tight" mitochondria in people with equatorial origins, it's predictable this uncoupler would have benefited these individuals to a significantly greater extent. Although it seems there are pharmacological interventions that can induce the production of uncoupling proteins, it's important to be mindful of their numerous adverse effects.

Then there's always the hope that a spontaneous genetic mutation could lead to the same benefits. In fact, a study published in the late 1990s found that nearly two-thirds of Japanese individuals who were over one hundred years old shared the same mutation within their mtDNA. This single letter change in the gene for a particular subunit of Complex I meant these lucky individuals were 50 percent more likely to live to one hundred years of age than those without the mutation (the general population). Further, the researchers discovered these individuals were 50 percent less likely to end up in the hospital for any reason at all in the latter half of their lives, and less likely to experience any age-related degenerative disease! An investigation into the effect of this genetic mutation showed that it resulted in a small reduction in the rate of free-radical leakage. Though not hugely significant at any given moment, this small reduction over the course of a lifetime ultimately led to a considerable benefit. This kind of evidence supports the Mitochondrial Theory of Aging and the premise that all age-related degenerative diseases can be addressed by targeting the health of mitochondria.

This beneficial genetic mutation is found only in Japan, though, so our only hope of replicating this study is the genetic modification of our own genes. However, there are obviously serious ethical and moral issues related to modification of genes, so we're left with a hunt for other alternatives. Genetic modification is not the only possibility.

If escaping electrons are responsible for free-radical production, the best way to prevent it would be to minimize the number of electrons passing through any given ETC (if there are relatively few electrons hanging out, they are less likely to leak to form free radicals). Having fewer electrons pass through any given ETC seems to be how the birds do it. The question is, how can humans minimize electron passage through ETCs? We can increase the number of ETCs per given mitochondrion to spread out the electron load; however, this number seems to be under genetic control, so it's probably not the lowest hanging fruit. Conversely, we could reduce the number of electrons, which is how caloric restriction works. Caloric restriction is currently the only proven method to extend life span in numerous mammals. (More on this in "Ketogenic Diets and Calorie Restriction," page 170.)

There are also numerous other ways we can optimize the health of our mitochondria. We can create more mitochondria, we can ensure electrons are removed from the complexes quickly (to make room for the next incoming electrons), we can dissipate the proton gradient as heat, and so on. The question is, how do we do these things?

We know that most cells generate 60–70 percent of all their energy from the metabolism of fatty acids. However, without adequate supply of nutrients such as L-carnitine to transport fatty acids into the mitochondria (and also remove toxic metabolites), cellular energy production will be inefficient—which I've already established as being the start of the end. The same is true for CoQ10. Stephen Sinatra, a cardiologist and one of the leaders in the field of metabolic cardiology, has tracked CoQ10 levels in hundreds of patients over twenty years and has found that levels of this critical component of the ETC are dangerously low in many more people than initially thought. Further, statin medications (the best-selling drugs prescribed in order to lower cholesterol levels) block the body's synthesis of CoQ10, so as more people are placed on these controversial drugs, we can logically predict that major deficiencies of CoQ10 will become more common. As previously discussed, other drugs such as beta-blockers, hypo-glycemic (diabetic) medications, and tricyclic antidepressants can further depress CoQ10 levels. This is all on top of the natural decline in CoQ10 production by the body as we age.

In addition, vegetarians and vegans, in most cases, don't get enough CoQ10 and L-carnitine through their diets because the primary dietary sources of numerous "mitochondrial nutrients" are meat. (We consume the animals' mitochondria and its components when we eat meat. Plants have chloroplasts that are similar to mitochondria, but they are not quite the same.) Table 3.1 lists the nutrients required for mitochondrial components.

You'll soon see how other nutrients, such as D-ribose, fit into this picture and why they are so critical for those affected by health conditions with decreased mitochondrial energy production. I'll also discuss the importance of magnesium—a popular mineral supplement—because a magnesium ion is almost always attached to ATP in the cells (it reduces the electrical charge of ATP and helps it move around in the cell). A

TABLE 3.1. Nutrients Required for
Mitochondrial Components

MITOCHONDRIAL COMPONENT	NUTRIENTS REQUIRED
TCA cycle	Thiamine (B_1) Riboflavin (B_2) Niacin (B_3) Pantothenic acid (B_5) Iron Sulphur Magnesium Manganese Cysteine Alpha-lipoic acid
Heme (required for elements in the TCA cycle and ETC)	Zinc Riboflavin (B_2) Pyridoxine (B_6) Iron Copper
Synthesis of L-carnitine	Vitamin C (or L-carnitine itself)
Pyruvate dehydrogenase	Thiamine (B_1) Riboflavin (B_2) Niacin (B_3) Pantothenic acid (B_5) Alpha-lipoic acid
Electron transport chain	Riboflavin (B_2) Iron Sulphur Copper Coenzyme Q10

lot of what I'll discuss comes from Sinatra's teachings and the field of metabolic cardiology. I will address questions such as why heart disease patients feel worse for several days following a stress test. In such a test, a patient is placed on a treadmill to increase the demand for oxygen and create a state of temporary *hypoxia* (lack of oxygen relative to demand);

when the treadmill stops and demand for oxygen returns to "normal," hypoxia should resolve. But why, then, do patients experience fatigue, weakness, and shortness of breath for several days afterward? As you might have guessed, it all relates to mitochondrial energy!

What follows is a discussion of the various nutritional and lifestyle factors that can boost energy production and improve mitochondrial health. Of course, this is by no means an exhaustive list; in fact, I was forced to minimize the breadth of nutrients and therapies included in order to publish this book in a timely fashion (and it still took over two years to research).

D-Ribose

In the 1940s and 1950s, research showed that D-ribose, a simple five-carbon sugar, was the primary intermediate in an important metabolic pathway called the *pentose phosphate pathway*. Prior to this discovery, this unique sugar was thought to only be used as a structural component of the genetic molecules DNA and RNA.

While D-ribose is important for energy synthesis (as a structural component of ATP), it wasn't until the 1970s that researchers discovered that with supplemental D-ribose, when given prior to or immediately after ischemia in the heart, energy-deficient hearts could recover their cellular energy levels.

In 1991, the first clinical study of D-ribose in cardiology was published. The researchers theorized that parts of the heart affected by ischemia and hypoxia were just hibernating and not actually dead; they were lying dormant and conserving energy until they had enough blood flow and oxygen to ramp up energy production and function properly once again.

The purpose of this study was to help doctors determine what areas of the heart needed to be supplied with a new blood supply through bypass surgery. If a segment was just hibernating, a surgeon could run a new blood vessel to that area. However, if an area was dead, there was no need to reroute blood to the dead zone.

What they discovered had repercussions that were significant not only for those with ischemia and hypoxia but also for the general population. By administering D-ribose and replenishing the purine pool and energy reserves, the researchers found they could "wake up" dormant sections of the heart, proving their hypothesis.

Since then, many more studies have been conducted, and the benefits of D-ribose supplementation include improved recovery from cardiac surgery, improved function of the heart in congestive heart failure, restored energy to depleted skeletal muscles, and others.

Restored energy to depleted skeletal muscles is important because while D-ribose, at this time, is underutilized for cardiology, it has gained significant traction among athletes. Between 2002 and 2004, significant studies were conducted that showed D-ribose supplementation resulted in a lower heart rate for a set amount of work on a stationary bike, improved diastolic function in the heart, increased exercise tolerance, and accelerated recovery of the energy pool of stressed skeletal muscles.

Athletic Heart Syndrome

These results from studies of D-ribose are highly significant because they give us a possible explanation for the diagnosis of an "athletic heart." Athletic heart syndrome, also known as *athlete's heart* or *athletic bradycardia*, is a condition commonly seen in sports medicine, in which the human heart is enlarged, and the resting heart rate is lower than normal. It is caused by significant amounts of aerobic exercise performed over a period of at least several months.

Athlete's heart is common in athletes who routinely exercise more than an hour a day, and occurs primarily in endurance athletes, though it can occasionally arise in heavy weight trainers. While the condition is generally believed to be benign, it might sometimes be hard to distinguish from other serious medical conditions, and it has caused the sudden cardiac death of many highly trained and seemingly healthy athletes.

Why? Let's think back to the process here. Remember that if ATP cannot be generated fast enough, two ADPs will combine to produce ATP and AMP. This AMP will be broken down and eliminated from the cell, reducing the purine pool. It takes significant time for the purine pool to recover naturally, but instead of resting and allowing the heart to recover, the athlete goes out the next day (or even later that same day) and does more exercise, which further depletes the purine pool.

As the energy pool gets minimized by successive bouts of exercise with little chance to recover, the heart starts to enlarge (referred to as *hypertrophy*) to increase the muscle mass to compensate for its inefficiency.

Eventually, "the straw that breaks the camel's back" is not any one particular intense bout of exercise (although we've seen many cases of healthy athletes experiencing a sudden cardiac arrest during an intense event such as a marathon), but just another regular exercise session. The threshold is breached and the heart can no longer compensate—through any mechanism—for the energy demands placed on it by the physical activity. The heart stops—not because of a heart attack in the traditional sense (a blockage cutting off blood flow)—but because the heart just simply runs out of steam. While athletic heart syndrome is typically thought of as a nonpathological condition, it's plausible, based on bioenergetic chemistry, that pathological effects, and even death, might occur in extreme situations.

D-ribose is likely one of the most important nutrients for elite athletes who want to maximize the benefits of training but minimize its health risks. It is possible to consume D-ribose naturally from some foods (such as milk and dairy products, eggs, and mushrooms), but you can't get enough from these food sources to realize the benefits seen in the above-mentioned D-ribose studies.

Cardiovascular Diseases

Essentially everyone with a cardiovascular disease will have some degree of energy deficiency. The heart is one of the most metabolically active tissues in the body, and because its energy production is almost entirely aerobically produced through oxidative phosphorylation, it requires a large and constant supply of oxygenated blood. Perhaps this is why the heart is the first organ blood flows to after picking up fresh oxygen from the lungs.

The heart's energy demand makes it particularly vulnerable to ischemia and hypoxia, and although it does have various mechanisms in place to help maintain its energy production when faced with oxygen deprivation, in reality, these quickly run dry—literally in seconds (which is the reason why every second counts during a heart attack).

Although ATP levels have been shown to decrease by up to 30 percent in failing hearts, the loss of the purine pool is hard to detect until the heart's function is severely affected.

Due, in part, to reduced oxygen levels and resulting loss of mitochondria (there is no need for mitochondria if there is no oxygen), the heart shifts energy metabolism to the less efficient pathway of glycolysis. Not only does this result in lactic acid buildup, but with reduced energy efficiency, there is a progressive loss in contractility. The heart tries to compensate by enlarging its size, and this in turn worsens the ejection fraction and diastolic function, which in turn further deprives the heart of oxygenated blood. It's a vicious cycle that continues unless nutritional intervention takes place.

The same situation is true for those undergoing medical intervention. After heart surgery or after a *clot busting* therapy (in cases of a heart attack or stroke), there is a rush of freshly oxygenated blood. However, during the preceding ischemia, the purine energy pool was significantly reduced and the electrons backed up in the ETC. With reperfusion, we have lots of oxygen, but not enough mitochondria or enough ETC within each mitochondrion. Further, with all the excess electrons primed for free-radical production, introducing oxygen-rich blood is a prescription for disaster. The result is a burst of superoxide free radicals (produced within the remaining mitochondria), the opening of the mPTP, and eventually the death of mitochondria and ultimately the cells. Sound familiar? It should; this is *ischemia-reperfusion injury*, or IRI, which we already discussed ("The Role of Mitochondria in the Nervous System, Brain, and Cognitive Health," page 75).

D-ribose is finding utility and acceptance in cardiac surgery; the heart is one of the organs that respond most favorably to D-ribose supplementation. Supporting the heart's ability to preserve and rebuild its energy pool by supplementing with D-ribose is one of the first steps in restoring energy efficiency in any cardiovascular condition. Studies have shown it is effective in improving cellular energetics in congestive heart failure, coronary artery disease, and angina.

Fibromyalgia

Discussed earlier, fibromyalgia is a common chronic syndrome in which a person has long-term, body-wide pain and tenderness in the joints, muscles, tendons, and other soft tissues. Many times, patients become so sore, weak, and fatigued that they are not able to do basic tasks, and this can be compounded by sleep problems, headaches, depression, and anxiety.

Research has revealed that in fibromyalgia patients, the lining of the *capillaries* (the tiny vessels that supply blood and oxygen to the muscles) becomes thickened. When this happens, oxygen cannot cross the blood-tissue barrier, and without enough oxygen to adequately supply the tissues, localized ischemia develops and drains the energy pool in the affected muscles. Without oxygen, the cells shift energy production from oxidative phosphorylation to anaerobic glycolysis. This results in lactic acid production and buildup, which aggravate symptoms of severe pain, muscle stiffness, soreness, and overwhelming fatigue. Also, because muscle relaxation takes more ATP than muscle contraction, the cell sustains a contraction and keeps the muscle tense.

Of course, there is more to the story. For example, the sustained increase in intracellular calcium during a contraction also causes potassium ions to rush out of the cell, activating pain receptors. Regardless, the point is that D-ribose administration in these patients would help rebuild the cellular energy pool, allowing the calcium pumps to work better, which would help manage the cells' calcium load, reducing the outflow of potassium ions and resulting pain, and relaxing the muscle. Many fibromyalgia patients who use D-ribose to support cellular bioenergetics report that they are able to become involved in the normal activities of daily life again.

Supplementing with D-Ribose

D-ribose is naturally present in food, but dietary intake is insufficient to provide any meaningful impact on purine pools, especially in cases of chronic illnesses. Our main source of D-ribose is our body's own production, which occurs in every cell of the body, beginning with glucose through the pentose phosphate pathway. However, because this pathway occurs slowly, the best way to quickly replenish D-ribose is to supplement. When administered, a whopping 97 percent (approximately) is absorbed into the blood, and eventually moves into tissues without any difficulty.

Once in the cells, D-ribose is used by the body to synthesize and salvage the energy pool, to produce RNA and DNA, and to manufacture other critical molecules used by the cell. Of all the naturally occurring sugars found in nature, D-ribose is the only sugar that functions in these essential metabolic processes.

While there is technically no such thing as a D-ribose "deficiency" in the traditional sense, it's obvious that in certain circumstances, there is a deficiency relative to the demand or speed at which we need it to be available.

For example, ischemic hearts can lose up to 50 percent of their energy pool. Assuming blood flow and oxygen supply are restored, it might take up to ten days for the heart to rebuild its energy pool naturally and restore diastolic function—and this assumes the heart is allowed to rest!

Without supplemental D-ribose, the heart is forced to create it from glucose (again, through the pentose phosphate pathway). However, the problem is that under ischemic conditions when oxygen is in short supply, the mitochondria cannot produce ATP through oxidative phosphorylation, and the cell must rely more heavily on anaerobic metabolism or glycolysis, which uses glucose. Glycolysis is great because it's fast, but the process needs a constant supply of glucose to ensure quick energy turnover. The downside of this is that the cell does not want to sacrifice or donate any glucose to the pentose phosphate pathway to produce D-ribose, making recovery in the absence of surgical or nutritional intervention highly unlikely. When supplemental D-ribose is given, the energy pool and diastolic function can return to normal within one or two days!

Who Should Supplement

Depletion of energy pools can happen for many reasons and can also be highly variable between people, so it's not easy to generalize who should supplement with D-ribose. For example, any type of exercise can cause energy pools to be depleted, but for highly trained athletes, it might take hours of exercise, while it takes only a few minutes for sedentary people to reach the same level of depletion. One generalization we can make, however, is that physical activity depletes energy pools, and D-ribose supplementation becomes progressively important with frequency and intensity of physical activity.

Further, CoQ10 levels naturally decrease as we age (discussed in "Coenzyme Q10," page 144), with related symptoms typically presenting starting in our forties. CoQ10 is an essential component of oxidative phosphorylation, and those in their forties and beyond typically show signs of mitochondrial dysfunction, making the cells rely more on glycolysis, tying up the available glucose for energy production. Therefore, without glucose readily available for the pentose phosphate pathway, supplemental D-ribose becomes increasingly important as we age.

D-ribose is also useful—and sometimes critical—for those using a ketogenic diet (discussed in more detail in "Ketogenic Diets and Calorie Restriction," page 170) or those who have dramatically reduced carbohydrate intake. Certain drugs can increase the need for D-ribose. For example, drugs that help the heart contract more forcefully will eventually deplete the heart's energy pool. Those with fibromyalgia, as we already discussed, as well as those with a host of other health conditions, will also benefit from supplementation.

How to Supplement with D-ribose

Any amount of D-ribose given to energy-depleted cells will help. Even doses as low as 500 milligrams could be beneficial, although likely are not nearly enough to make a real improvement in health. Standard dosages range from 3 to 5 grams per day.

For healthy people and athletes, a dose taken before exercise helps the cell with the process of purine salvage as purines are broken down. A dose after exercise helps speed the *de novo* process to aid recovery. For people with a chronic health condition, an adequate dose will usually result in symptom improvement, typically within a few days. If a standard dose doesn't help, increase the dose until an effect is noticed (e.g., relief of symptoms). D-ribose is safe even at large doses, and many clinical trials have studied amounts ranging from 10 to 15 grams per day, with one study of McArdles's disease using 60 grams per day! If you think this is an extremely high load of sugar, note that D-ribose doesn't impact blood glucose or insulin levels like glucose does. It's completely safe for diabetics, even at these high doses.

Since it is blood that delivers D-ribose to the tissues in need, for some with blood flow issues, the dose might need to be substantial in order for the D-ribose to work its way through areas of reduced blood flow in sufficient quantities to have a noticeable impact on symptoms. Further, because the energy pool will continue to be drained, it's important to ensure continual supply of D-ribose to the cells. This means daily supplementation is usually in order, though more so for those with chronic health conditions than for healthy individuals.

Pyrroloquinoline Quinone (PQQ)

Traditionally, it was believed that generating new mitochondria (mitochondrial biogenesis) could only occur as a result of strenuous exercise or extreme calorie restriction, which is why research on PQQ (*pyrroloquinoline quinone*) is so exciting. Early in 2010, researchers found PQQ not only protected mitochondria from oxidative damage, it also stimulated the growth of new mitochondria!

Mechanisms and Functions in Humans and Animals

A number of physiological properties have been attributed to PQQ, ranging from classic water-soluble vitamin and cofactor functions to protection of nerve cells, promotion of nerve growth, and mitochondrial biogenesis. While a role as a vitamin in animal or human nutrition might (or not) be possible pending further research, at this time, similar to other related compounds, there is strong evidence PQQ might play an important role in pathways important to cell signaling. The importance of PQQ to mammalian health is evident when it is omitted from chemically defined diets, resulting in a wide range of systemic responses, including growth impairment, compromised immune responsiveness, and abnormal reproductive performance in experimental mouse and rat models. Further, varying PQQ levels in diets cause modulation in mitochondrial content, alter lipid metabolism, and reverse the negative effects of Complex I inhibitors.

Under appropriate conditions, PQQ is capable of catalyzing *continuous redox cycling* (the ability to catalyze repeated oxidation and reduction reactions), which is a novel chemical property in many respects. For example, in chemical assays, PQQ's stability renders it capable of carrying out thousands of redox catalytic cycles, whereas other bioactive quinones capable of redox cycling (e.g., epicatechin in green tea) tend to self-oxidize or form polymers (e.g., tannins), rendering them useless in further redox reactions. PQQ and its principal derivative, IPQ, are widely distributed in animal and plant tissues ranging from pico- to nanomolar concentrations.

From an evolutionary perspective, current evidence suggests PQQ is a component of interstellar dust, and since it's been postulated that strong redox catalysts were required to trigger the earliest chemical evolutionary steps, the extraterrestrial origin of PQQ raises the question of its evolutionary importance to simpler life forms. This theory is especially interesting when you consider PQQ's wide range of chemical properties, such as a redox catalyst and the ability to modify amino acids (e.g., oxidative deamination reactions). Could PQQ be our common point of origin with life in other areas of the galaxy?

Mitochondrial Biogenesis and PQQ

Improvements in mitochondrial energy production are potentially important for treating a number of health issues, as previously discussed, so increasing the number of mitochondria in any given cell can have far-reaching benefits ranging from increased longevity to improved energy utilization and protection from free radicals.

Many mitochondrial-related events are regulated by peroxisome proliferator-activated receptor-gamma coactivator-1 alpha (PGC-1alpha) and nuclear respiratory factors. PGC-1alpha is a transcriptional coactivator that regulates genes involved in energy metabolism. An interaction with this protein and its resulting association with multiple transcription factors can provide a direct link between an external physiological stimulus (such as PQQ) and the regulation of mitochondrial biogenesis. Indeed, such interactions have recently been reported.

PGC-1alpha is also a major factor that regulates muscle fiber type and appears to be involved in the control of blood pressure, the regulation of cellular cholesterol homeostasis, and the development of obesity. Moreover, PGC-1alpha is associated with a reduction in free radicals and protection against various mitochondrial toxins.

In addition to interacting with PGC-1alpha, PQQ can also reduce cancer risk by mechanisms separate from mitochondrial biogenesis. For example, PQQ has been shown to affect the activity of *ras* (a gene that can potentially cause cancers). PQQ administration has been shown to activate other transcription factors, such as nuclear respiratory factors (NRF 1 and 2) and mitochondrial transcription factors (e.g., Tfam), which lead to increased mitochondrial biogenesis.

PQQ also has another beneficial effect on mitochondria. It appears that PQQ might be an essential cofactor in one of the many protein subunits that make up Complex I of the ETC. With the bulk of endogenous free radicals produced at Complex I, you can see why having an abundance of PQQ is important to mitochondrial health.

Given its powerful effects on mitochondrial biogenesis, it makes sense that PQQ would result in powerful effects on various health conditions. Studies in both animals and humans have shown that PQQ improves reproduction, early development, growth, and immune function. It has protected nerve cells from degeneration and damage, and even promoted the growth of nerve cells and helped form new synapses (connections) between nerve cells (important in the brain for memory). For cardiovascular health, it has reduced the damage from ischemia-reperfusion injury, heart attacks, and stroke.

Is PQQ the Latest New Vitamin?

In 2003, Japanese scientists made a significant announcement in the prestigious journal *Nature*: They had discovered direct molecular evidence that PQQ was in fact a previously unidentified B vitamin. A vitamin, by definition, is a compound that human bodies can't make (we *must* get it from dietary sources) and is an absolute necessity to carry out at least one essential biochemical function.

The first and most obvious evidence of its vitamin role came from animal studies, where researchers found that in mice, eating a

PQQ-free diet led to impaired reproductive and immune function. Further, the growth of the mice was impaired and their skin became thin and fragile. Their offspring were less likely to survive the first few days after birth. However, and most importantly, PQQ-deficient mice had 30–40 percent fewer mitochondria. Further, the mitochondria that they did have were abnormally small and did not appear to function properly. None of these signs and symptoms occurred in the mice that ate the same diet supplemented with PQQ.

However, proving a chemical compound is a vitamin is a complex task, and it was a serendipitous discovery by the Japanese researchers that gave the most direct evidence of PQQ's role as a vitamin. The discovery was made while the researchers were investigating whether *bipolar disorder* (previously known as manic depression) involved abnormalities in mitochondrial calcium transport. They were looking for the genes responsible for building the proteins that control the transport of calcium into the mitochondria. Although their work on bipolar disorder wasn't completed as planned, what they discovered was how PQQ is involved in activating a key enzyme in collagen production (the basic structural protein in skin, bone, and connective tissue)—a connection previously suggested by the brittle skin and dysfunctional connective tissue of PQQ-deficient animals. This same PQQ-dependent enzyme is found in humans.

More recent evidence has thrown into question whether PQQ is a vitamin, but research is ongoing, and at this time it seems a little too early to make a conclusion either way.

Other Roles of PQQ

Studies suggest that PQQ might also (1) have anti-inflammatory effects, (2) be an effective neuroprotectant (reducing brain damage during simulated stroke and protecting brain cells against "excitotoxic" overstimulation), and (3) be a stimulator of *nerve growth factor* (NGF, a key protein involved in the growth and survival of nerve cells).

Related to this is cognitive function. As we discussed, the brain uses an incredible amount of energy, and it is utterly dependent on the mitochondria for fuel. A double-blind, randomized, placebo-controlled human clinical trial found that 20 milligrams

TABLE 3.2. PQQ Content of Various Foods

Food	PQQ (MICROGRAM PER KILOGRAM OF FOOD)
FRUITS	
Kiwifruit	27
Papaya	27
Tomato	9
Orange	7
Apple	6
VEGETABLES	
Parsley	34
Green tea	30
Green pepper	28
Carrot	17
Cabbage	16
Celery	6
LEGUMES	
Tofu	24
Broad beans	18
Soybeans	9
FERMENTED FOODS	
Cocoa powder	800
Natto (fermented soybeans)	61
ANIMAL PRODUCTS	
Human breast milk	140–180
Egg yolk	7
Egg white	4
Cow's milk	3

of oral PQQ taken daily improved short-term memory, attention and concentration, information identification, and processing ability in healthy adults. Effects were greatly enhanced with the addition of CoQ10 supplementation, which makes sense because Complex I transfers its electrons to CoQ10 in the ETC. If PQQ is stimulating the production of more mitochondria, and there is a corresponding increase in the number of Complex I units, we need to increase the concentration of CoQ10 to ensure that all the ETCs in all those mitochondria are passing their electrons down the chain uninhibited.

Dark Chocolate

Table 3.2 shows the high content of PQQ in cocoa powder. Perhaps this is one reason for the many health benefits associated with chocolate consumption. Of course, there are plenty of other health-giving compounds in chocolate (such as flavonols, theobromine, and epicatechin), but we should definitely not discount the high concentration of PQQ.

In fact, research into the health benefits of chocolate compounds has shown cardiovascular and neurological/cognitive health benefits, improved exercise performance and endurance, and even weight-loss benefits. When you consider the benefits of mitochondrial biogenesis, it's logical to see how PQQ could be the link that explains these effects. More research should be done to clarify PQQ's contribution to the health benefits of chocolate, but in the meantime, moderate consumption of dark chocolate would seem to be a great way to treat yourself. Personally, I usually overdo it with this therapy. Feed me chocolate and I'll do whatever you say.

Coenzyme Q10

Coenzyme Q10 (CoQ10) is an antioxidant, a membrane stabilizer, and a vital component in the mitochondrial ETC. It also regulates gene expression and apoptosis; is an essential cofactor of uncoupling proteins and permeability transition pores; and has anti-inflammatory, redox modulatory, and neuroprotective effects.

CoQ10 is a vitamin-like molecule that is naturally present in just about every single cell in our body. Like vitamins, it is absolutely essential to life. However, since our bodies produce CoQ10, it's not technically a vitamin. In order to produce CoQ10, the cell needs an amino acid called tyrosine, at least eight different vitamins, and several trace minerals. A deficiency of any of these impairs the cells' ability to produce CoQ10.

Despite the fact that we can produce it, CoQ10 becomes more vitamin-like as we age because we produce less and less with advancing age (with the body slowing its production starting in our late twenties or early thirties). Many believe this is by design, because by our late twenties, our prime reproductive years are behind us, and as we raise our children to take our spot on this finite planet, reduction of CoQ10 is Nature's way of preparing us for our eventual exit. Considering how important mitochondria and oxidative phosphorylation are to our health and longevity, by reducing the production of this one molecule, the body can start winding things down to make room and free up resources for the next generation.

Although we do get small amounts of CoQ10 from our food, it's literally a few milligrams daily—not nearly enough for our bodies to benefit clinically, and supplementation becomes increasingly important the older we get. Unfortunately, absorption of this rather large fat-soluble molecule is challenging, which is a main factor limiting its therapeutic use.

Research has shown that oil-based formulations (typically softgels) are much better absorbed, and water-dispersible liposomal or pre-emulsified formulations are even better. *Ubiquinol* (reduced CoQ10) seems to offer much better absorption than *ubiquinone* (oxidized CoQ10), and water-soluble (*solubilized*) ubiquinol is even better absorbed.

Many familiar with CoQ10 also know it as an antioxidant, and its antioxidant properties are directly related to its primary function in the ETC, where it participates in redox (reduction-oxidation) reactions. When it picks up an electron from Complex I or II, CoQ10 becomes reduced. For this reason, CoQ10 is arguably the single most important nutrient for mitochondrial health. If the bulk of free radicals are formed at Complex I, we could probably guess that the next step in the chain would be the bottleneck. In fact, this has not only been confirmed by many studies, but therapeutically, CoQ10 supplementation has been shown to dramatically improve the status of patients with all sorts of health conditions. In a sense, CoQ10 supplementation essentially rescues failing bioenergetics, and targets *the* main site of free-radical production.

CoQ10 can even take free radicals and put them (or more specifically, their electrons) to good use, as it can bring those rogue electrons back into the ETC for energy production. But even more importantly, this antioxidant activity will help prevent the associated damage typically caused by free radicals, by protecting mtDNA, membranes, and other peptides and enzymes.

Although 80 percent of CoQ10 is found in mitochondria, its presence in microsomes, Golgi apparatus, and plasma membranes indicates its importance as an endogenously produced lipid-phase antioxidant. Even in mitochondria, up to a third of CoQ10 is bound to mitochondrial membrane proteins, apparently to serve primarily as an antioxidant. Long-lived mammalian species show a greater proportion of mitochondrial membrane-associated CoQ10 than short-lived species.

Congestive Heart Failure and CoQ10

Congestive heart failure (CHF) and dilated cardiomyopathy are both conditions where the heart muscle is so weak that it can't contract and pump blood effectively, which causes the blood to back up or become "congested," especially in the legs and the lungs. This congestion sets up a chain reaction because the blood can't be oxygenated properly (inefficient flow to the lungs), and without oxygen as the final acceptor of electrons at Complex IV, the ETC backs up and starts to spill free radicals, aggravating that vicious cycle that you're probably sick of hearing about by now.

One of the greatest natural tools the medical community has against CHF is CoQ10, and its benefit to CHF is due to multiple actions. Medical literature is overflowing with studies showing the benefits of CoQ10 for CHF, but the dosages typically used in those studies (especially the earlier studies) were too low to provide a benefit —compounded by the fact that early formulations were difficult to absorb. New research using higher doses and highly absorbable formulations is proving CoQ10 could very well be one of the single most important nutrients for mitochondrial health in CHF.

Hypertension

CoQ10's effectiveness in lowering blood pressure has been known since the 1970s. It does so in a number of ways. First, as an antioxidant, it can neutralize *peroxynitrite free radicals*. Peroxynitrites are generated from an important molecule called *nitric oxide*. The benefit of nitric oxide is that it can help dilate (relax) blood vessels and reduce the platelet "stickiness," which ultimately lowers blood pressure. In fact, many medical therapies for blood pressure, both conventional and traditional, target the nitric oxide pathways. Unfortunately, as nothing is 100 percent good and nothing is 100 percent bad, an overabundance of nitric oxide results in peroxynitrites that can damage the blood vessels (both the endothelial cells that line the blood vessels and the smooth muscle cells surrounding them). Thankfully, in normal healthy individuals, over 90 percent of the circulating CoQ10 in the blood is found as ubiquinol, the powerful antioxidant form that can help minimize the damage and maximize the benefits of nitric oxide to the cardiovascular system.

Second, it can prevent the oxidation of LDL (bad cholesterol), which, when oxidized, can lead to plaque buildup and hardening of the blood vessels (called *atherosclerosis*). As long as LDL is not oxidized, it's actually not a bad thing (contrary to what many people think based on conventional medicine).

Third, remember that muscle relaxation takes more ATP than muscle contraction (remember our discussion of rigor mortis). ATP is critical for relaxation and this becomes especially important for the smooth muscles that line the blood vessels. Without enough energy, these muscles remain tenser than they should, increasing blood pressure. By supplying CoQ10 and improving the energy efficiency of the mitochondria, muscles have the ATP they need to relax, thereby normalizing blood pressure. I specifically say "normalize" as opposed to "lower" blood pressure, because clinical trials have shown that CoQ10 can lower high blood pressure, but will not lower normal or low blood pressures. I should quickly mention, however, that those taking blood pressure medications should know that additive effects have been reported, and supplementing with CoQ10 without adjusting medications sometimes can result in blood pressure that is too low. While most people with high blood pressure who supplement with CoQ10 are able to lower the dose of their medication (or even stop in some cases), it's always best to discuss how to take advantage of this effect with your health care practitioner.

Lastly, research has shown that CoQ10 can indirectly influence blood vessel function by improving blood sugar control. High blood sugar increases oxidative stress, which, as described previously, damages the blood vessels and makes them stiffer.

Heart Protection during Cardiac Surgery

There are three main types of heart surgery, all intended to restore blood flow to an ischemic heart. Regardless of the surgical procedure, the result is that blood flow is restored. Sounds great, right? Wrong. Remember reperfusion injury, which we discussed earlier? When reperfusion (restoring blood flow) allows oxygenated blood to finally reach the areas of the heart that were hypoxic/ischemic for a long period of time, the cells that were once starved of oxygen have a fresh supply of highly oxygenated blood delivered directly to them and the result is a burst in superoxide free radicals.

Reperfusion injury is one of the main side effects of these life-saving surgical procedures. However, as a powerful antioxidant, CoQ10 minimizes the damage caused by these superoxide free radicals, thereby improving the results of heart surgery and speeding recovery.

CoQ10 as Adjunct to Statin Therapy

Statins are a class of prescription drugs that are widely used to treat high cholesterol levels. In fact, they are among the most overprescribed drugs in the world. This highly controversial group of drugs lower cholesterol by blocking the key enzyme (HMG CoA reductase) in our bodies' own production of cholesterol. This enzyme is targeted because approximately 80 percent of the cholesterol in our bodies is made internally, not obtained from diet. This same enzyme, however, is involved in making CoQ10 (and also vitamin D, all the sex hormones, and so on); many of the adverse side effects associated with statins (such as muscle pain and muscle damage) are theorized to be caused by an "induced deficiency" of CoQ10.

Patients performing endurance exercise while taking a statin have had significantly more muscle damage compared with patients not taking a statin, which further suggests that the side effects of statins are due to this induced CoQ10 deficiency (muscles require high amounts of ATP and, therefore, CoQ10 during physical activity). There are also at least two randomized, controlled trials that showed significant lowering of the severity of muscle pain with the use of CoQ10 in patients taking statins.

Thus, it is reasonable to recommend that almost everyone taking a statin medication, especially if experiencing muscle pains, should supplement with CoQ10. Many progressive cardiologists and pharmacists are now recommending CoQ10 anytime a statin prescription is filled. In fact, the rationale for using CoQ10 to treat statin-induced muscle pain was so overwhelming that Merck & Co, Inc., decided to pursue a patent for a CoQ10-statin combination product. Merck & Co, Inc., was eventually issued two patents for this combination for counteracting the statin-associated disease of the muscles.

Other Drug Interactions

CoQ10 is likely one of the safest nutrients to take with pharmaceutical drugs, which is great because the people who need it the most are likely on a cocktail of different drugs. In fact, there are certain instances where CoQ10 is highly recommended with certain drugs, but other situations where some caution must be exercised.

Beta-blockers are a group of drugs, typically prescribed for hypertension and arrhythmias, that have been shown to deplete CoQ10 levels—meaning for those who take beta-blockers, CoQ10 supplementation is recommended (as it is in statin therapy). In fact, CoQ10, when given concomitantly with beta-blockers, was shown to reduce the fatigue that is usually induced by these drugs. However, people taking beta-blockers should be aware that there could be possible additive effects, as there sometimes are when CoQ10 is taken with blood pressure medications.

The main drug that is of concern when supplementing with CoQ10 is *warfarin*, a blood thinner. Until recently, warfarin was considered first-line therapy for *atrial fibrillation* (fluttering of the heart—when the heart flutters and creates turbulence in blood flow, there is a greater risk of clots forming). Warfarin is a vitamin K antagonist, and it "thins" the blood by blocking vitamin K's ability to activate clotting factors (which is why anyone taking warfarin must watch their intake of vitamin K, even from foods). The chemical structure of CoQ10 and vitamin K is very similar (they're both *quinones*), so there is a potential for CoQ10 to decrease the effectiveness of warfarin. However, CoQ10 will not have a negative impact on other classes of blood thinners—only warfarin—and CoQ10 will not "thicken" blood (knowledge of the biochemistry of clotting is required to understand this, but that is beyond the scope of this book).

On the other hand, CoQ10 can have antiplatelet action, similar to that of another class of blood thinners called "antiplatelet agents." It seems that CoQ10 can reduce the "stickiness" of platelets, and can help prevent clots from forming.

Neurodegenerative Diseases

Research has also shown that CoQ10 might help neurodegenerative diseases, such as Huntington's disease and *amyotrophic lateral sclerosis* (ALS, or Lou Gehrig's disease). Additional research findings suggest that CoQ10 supplementation might also help various forms of *ataxia* (reduced control over voluntary movement), particularly cases that show a decreased level of CoQ10 in their muscles. For example, scientists found marked improvements in mitochondrial defects among Friedreich's ataxia patients, which involve a deficiency of a mitochondrial protein called frataxin.

The bioenergetic effect of oral CoQ10 supplements in people with Parkinson's disease has also been studied. Researchers found that CoQ10 restored the depressed activity of Complex I to approximately normal levels.

Restoring the Stress-Recovery Response

The capacity to respond to stress declines with age. Younger patients, for example, recover faster from a heart attack or heart surgery than older patients. Studies have demonstrated that tissues from older patients show significantly less recovery following the stresses of hypoxia and ischemia. Further, there is a significant correlation between the integrity of mtDNA and the ability of tissues to recover from stress. Not surprisingly, CoQ10 is able to minimize, even abolish, these differences. These studies link the Mitochondrial Theory of Aging to the stress response in heart tissue, and they demonstrate how CoQ10 can restore energy levels and stress recovery in the aged heart (to youthful levels).

Why Supplement with CoQ10?

As mentioned earlier, the production of CoQ10 within our cells requires numerous other nutrients, and this is a significant challenge, especially in a society where food quality and dietary choices are far from optimal. In his book *The Sinatra Solution*, Sinatra describes a study where hospitalized patients were on total intravenous nutrition without vitamin and mineral support, and the CoQ10 levels in their blood collapsed by 50 percent in just one week. Also consider that the older we get, the less efficient our digestive system becomes at extracting nutrients from our food, so this becomes a concern for all cognizant of their CoQ10 status.

Second, as we enter our thirties, we're naturally not producing as much CoQ10 as we did in our younger years, and our production continues to decline as the years pass us by. Even though this is a natural phenomenon, and while some would question why we would want to mess with Nature, I'm pretty sure few would willingly suffer through the host of degenerative diseases associated with dysfunctional mitochondria brought about by CoQ10 deficiency. I'm sure most readers would like to live their lives—whether long or short—in the best health possible. Thus, preventive use of CoQ10 could possibly start in our late thirties.

Third, there are numerous medications that can deplete CoQ10 levels. As I previously described, the most recognized of these are the statins. Also as mentioned, at least two clinical studies have shown that people who cannot tolerate statin medications due to their side effects can tolerate them when they are supplemented with CoQ10. This data is where I get a little puzzled. CoQ10 is essential for heart muscle function—we know this. The heart requires vast amounts of energy to repetitively contract and relax and pump blood twenty-four hours a day for the duration of a life. It seems illogical to prescribe a drug that, while effective in lowering cholesterol, can actually promote dysfunction in the heart muscle and set one up for a future cardiovascular disease.

It's also important to remember that CoQ10 is an important antioxidant that prevents oxidation of LDL cholesterol (predominantly as ubiquinol when found in the blood). If CoQ10 is one of the primary antioxidants that keep LDL from oxidation, and statins deplete CoQ10 levels, again, you can see the paradox in prescribing statins.

Studies have clearly shown that, if CoQ10 is made available to the cells, the cells will take it up, increasing their CoQ10 levels and levels of mitochondria as well. Further, studies have shown we can significantly increase our CoQ10 level when we take it as a supplement, as long as a high-quality and absorbable formulation is used.

Should you choose to supplement with this critical nutrient, the ideal dose for cardiovascular benefits would be one that raises blood levels above a minimum of 2.5 micrograms per milliliter, but ideally over 3.5 micrograms per milliliter. Unfortunately, taking regular blood tests to determine a suitable dose is not only inconvenient, it's not readily available through most labs. For neurological conditions such as Parkinson's, Huntington's, or Alzheimer's diseases, the dose will need to be significantly higher because the blood levels need to be saturated to force it past the blood-brain barrier, but blood levels needed for therapeutic benefits in these conditions haven't been determined at this point.

The next best way to determine a good dose is symptom relief. For example, someone experiencing high blood pressure can start with 100 milligrams daily for a few weeks. If blood pressure doesn't come down, increase the dose to 200 milligrams daily for another couple weeks, then 300 milligrams daily, and so on until the desired effects are seen. This process works well for most, but the problem with this approach is that it assumes a 100 percent success rate, and nothing—whether nutritional supplements, drugs, or surgery—has a 100 percent success rate. There could be some people who will continue to take higher and higher doses but who will continue to experience their disease symptoms.

Therefore, the easiest approach (although the least individualized) is to work in typical dosage ranges. For example, cardiovascular conditions are typically dosed between 200 and 600 milligrams per day. Doses for neurological conditions range from 600 to 3,000 milligrams per day (no, that's not a typo, and even that high dose was perfectly safe). However, large daily doses should be divided up into multiple smaller doses taken throughout the day and, unless a solubilized formulation is used, should be taken with food.

Once a therapeutic dose is achieved, it must be maintained or the symptoms will return. This is because our bodies won't miraculously start producing copious amounts of CoQ10 again, as we did in our teenage years. In fact, there's a possibility that as time goes on, the dosage might need to be increased (as the body continues to naturally decrease its own production).

Lastly, it's important to point out that there are big differences in the quality and effectiveness of CoQ10 formulations on the market. While I mentioned earlier that oil-based formulations are better absorbed, and that water-dispersible or pre-emulsified formulations are best, in general, ubiquinol seems to be the ideal form to take as a supplement.

L-Carnitine

L-carnitine (*levocarnitine*) is similar to CoQ10, in that our bodies produce a significant amount, but its production is also thought to decrease as we age, so it becomes more vitamin-like in our elder years. There is considerable positive research supporting L-carnitine's use supplementally in numerous health conditions.

Functions of L-Carnitine

L-carnitine is a naturally occurring compound found in all mammalian species. The most important biological function of L-carnitine is in the transport of long-chain fatty acids into the mitochondria for subsequent beta-oxidation (to produce ATP). For most of us, just about *all* of our dietary fatty acids are long-chain fatty acids. In order to transport these fatty acids into the mitochondria, L-carnitine attaches to them to form *acylcarnitine* derivatives (the "acyl" just means something is attached to the L-carnitine molecule).

The physiological importance of L-carnitine and its obligatory role in the mitochondrial metabolism of fatty acids have been clearly established; however, more recently, additional functions of the carnitine system have been described, including the removal of excess acyl groups from the body (an important detoxifying role) and the modulation of intracellular coenzyme A homeostasis (which is critical in the TCA cycle).

The concentrations of L-carnitine and its acylcarnitines are maintained within relatively narrow limits for normal biological functioning in their pivotal roles in fatty acid oxidation and maintenance of free coenzyme A (CoA) availability. The homeostasis of carnitine is multifaceted, with concentrations achieved and maintained by a combination of oral absorption, biosynthesis, carrier-mediated distribution into tissues, and extensive kidney reabsorption.

This role of L-carnitine in buffering the ratio of free CoA to acyl-CoA is a function that is particularly important under conditions of stress. Under normal conditions, short- and medium-chain acyl-CoA, formed as a result of various mitochondrial pathways, is further metabolized to generate free CoA. However, under abnormal conditions in which excess molecules of acyl-CoA are formed within the mitochondria, acyl-CoA can react with L-carnitine to form acylcarnitines, thereby freeing CoA for use in other mitochondrial reactions. This reversible exchange, in combination with the ability of the resultant acylcarnitine to cross the mitochondrial membrane, means that the intramitochondrial relationship between free CoA and acyl-CoA is reflected in the extramitochondrial ratio of acylcarnitine to L-carnitine, an indicator of mitochondrial metabolic health.

Various disorders of carnitine insufficiency have been described, but ultimately all result in impaired entry of fatty acids into the mitochondria and consequently disturbed lipid oxidation.

Mitochondrial Fatty Acid Oxidation

The mitochondria's preferred source of fuel is fatty acids. Remember, fatty acids are so dense in energy that approximately 60–70 percent of the total amount of ATP our bodies produce originates from fatty acids.

The mitochondrial metabolism of cytosolic fatty acids begins with the formation acyl-CoA (in this case the "acyl" is the long-chain fatty acid). This acyl-CoA then combines with carnitine to create acyl-carnitine and a free CoA. This acyl-carnitine is then able to cross the outer mitochondrial membrane into the intermembrane space. From there, acyl-carnitine needs the help of a specific transport enzyme embedded in the inner mitochondrial membrane, which works by exchanging a free carnitine from within the matrix with an acyl-carnitine from the intermembrane space. Once inside the matrix, the reactions are reversed; acyl-carnitine reacts with free CoA, to

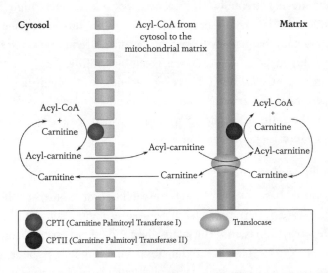

FIGURE 3.1. Schematic of L-carnitine's role in the transport of fatty acids (designated as an "acyl" group) across the inner mitochondrial membrane. Long-chain fatty acids cannot cross either the outer or inner mitochondrial membranes, so a transporter in the form of L-carnitine is needed. This diagram also depicts how coenzyme A (CoA) is intricately involved in the process, and illustrates how L-carnitine helps to manage CoA homeostasis.

form the corresponding acyl-CoA and free carnitine. Then the long-chain fatty acid (as acyl-CoA) can enter the fatty acid beta-oxidation pathway to produce acetyl-CoA, which ultimately produces ATP (see figure 3.1).

Given that neither acyl-CoA nor free fatty acids are able to move across the inner mitochondrial alone, the role of L-carnitine and carnitine acyltransferases in fatty acid metabolism is vital.

Lactic Acid Metabolism

Another important role for L-carnitine is clearing lactic acid buildup. As previously discussed, lactic acid, or lactate, is a by-product of *anaerobic metabolism* (when there isn't enough oxygen for oxidative phosphorylation, or when energy production needs to proceed superfast).

Many readers will have experienced lactic acid burn in their muscles during strenuous physical activity. Unfortunately, high levels of lactic acid damage tissues, evidenced by the muscle damage and resulting pain that usually follow the day after strenuous exercise.

In a study where one group was given L-carnitine, the rise in lactic acid in response to exercise was significantly lower than that in the control group. L-carnitine also helps speed recovery by helping to restore the ratio of lactate to pyruvate (which means less burning during exercise and less pain afterward).

Other Benefits and Uses

L-carnitine has also been well researched for numerous health conditions (e.g., peripheral vascular disease, angina, congestive heart failure, arrhythmias, infertility, fatty liver and other hepatic disorders, diabetes, exercise tolerance, weight loss), but ultimately, its benefits all related back to the biochemical roles discussed previously.

Dietary Intake and Absorption

Dietary L-carnitine intake is largely achieved via consumption of animal-based products, including red meats, poultry, fish, and dairy products, while negligible quantities are available from plant-derived foods. Given the broad range of nutritional choices, dietary

L-carnitine intake can vary considerably, with the standard omnivorous diet providing 6–15 millimoles per kilogram a day and the standard vegetarian diet providing less than 1 millimole per kilogram a day. However, despite substantial differences in L-carnitine intake, previous research has established that a vegetarian diet does not result in a significant deficit in the body's carnitine concentrations. In fact, on average, plasma L-carnitine, total carnitine, and estimated acylcarnitine concentrations for vegetarian adult subjects were only 10–20 percent lower (compared to adults consuming an omnivorous diet). On the other hand, urinary carnitine excretion for L-carnitine was 85–95 percent lower (and acylcarnitine was 40–50 percent lower) for vegetarians than for nonvegetarians. These findings indicate that compensatory mechanisms, including conservation by the kidneys in conjunction with biosynthesis, are effective in maintaining carnitine homeostasis when dietary L-carnitine intake is low.

The Body's Synthesis of L-Carnitine

While in omnivorous humans the majority of total body carnitine concentrations is achieved from dietary sources, approximately 25 percent comes from our own biosynthesis. On the other hand, strict vegetarians obtain very little L-carnitine from dietary sources and, consequently, in order to maintain carnitine homeostasis, biosynthesis accounts for up to 90 percent of total body carnitine concentrations in this group.

L-carnitine is synthesized from the amino acid precursors *lysine* and *methionine* (both essential amino acids), with lysine providing the carbon backbone and methionine acting as the methyl donor. The biosynthesis of L-carnitine involves other nutrients, including iron, vitamin C, oxygen, pyridoxal-5-phosphate (the biologically active form of vitamin B_6), and B_3 (as NAD^+).

Magnesium

Magnesium is likely one of the most underrated minerals and most people are just not consuming enough of it. One reason most are deficient in this mineral is that water softeners, while great for making your faucets shiny, has reduced the water's hardness by removing minerals such as magnesium. Further, high intake of calcium can reduce the absorption of dietary magnesium, and with the conventional medicine focus on calcium intake for bone health, we've seen a corresponding general decrease in magnesium levels. Then there's our rising caffeine intake, which increases the amount of magnesium we lose through urine, and our rising use of antacids and proton pump inhibitors—drugs that can reduce the absorption of magnesium. All these factors contribute to the alarming statistic that 70–80 percent of the developed world population is deficient in magnesium.

And that's only part of the picture! Regardless, it all equals bad news, because magnesium is a critical cofactor in over three hundred biochemical reactions in the body, including the production of ATP. Further, we've known since at least 1976 that mitochondria act as our intracellular magnesium stores, and a lot of the magnesium in the body is found bound to ATP, which helps stabilize ATP and make it usable by the body. In fact, when we talk about ATP in biology, we're actually talking about Mg-ATP; that's how important magnesium is!

Magnesium and Cardiovascular Diseases

Due to its role in energy production and metabolism, magnesium has countless benefits to just about every physiological system in the body. However, most people know it as a heart-healthy mineral because of its role in muscle relaxation. Remember that muscle contraction is initiated when calcium rushes into the cell. For muscles to relax, they not only require ATP (in which magnesium plays a role), but the enzymes involved in this relaxation process also require magnesium as a cofactor. Without magnesium, calcium cannot be removed from the muscle cell and the muscle remains in a contracted state. Magnesium has been labeled as "Nature's calcium channel blocker" (calcium channel blockers are a class of drugs commonly used to treat hypertension).

For the smooth muscles that surround the blood vessels, a magnesium deficiency means that they remain tenser than they should, a condition called *vasoconstriction*. Vasoconstriction can further aggravate any health condition because it restricts blood flow to a tissue and its cells. With less blood flow, there is less oxygen delivery. With less oxygen, oxidative phosphorylation in the mitochondria can't operate at top speed. Similarly, a deficiency of magnesium doesn't allow the heart to fully relax between contractions (we discussed this diastolic dysfunction previously in chapter 2, "The Basics of Cardiac Physiology," page 73).

Magnesium deficiency has been implicated in hypertension, ischemic heart disease, congestive heart failure, arrhythmias, angina, sudden cardiac death, atherosclerosis, mitral valve prolapse, cerebrovascular disease, and stroke. It's also been linked to preeclampsia and eclampsia, asthma, insulin resistance and diabetes, metabolic syndrome, osteoporosis, and even colon cancer.

Alpha-Lipoic Acid

Alpha-lipoic acid (ALA) is a molecule found in the mitochondria. Under optimal conditions, the body can manufacture sufficient amounts for its metabolic functions (it's a cofactor for the enzymes catalyzing the final stages of glycolysis, from which the resulting compounds can enter the TCA cycle).

However, additional ALA provided by supplements allows it to circulate in a "free" state, where it has the ability to function as both a water- and fat-soluble antioxidant. This is unique because most antioxidants are effective in only one area or the other. Vitamin C, for example, is usually restricted to the interior *cytosolic* (watery) compartment of cells, while vitamin E works at the level of the fatty cell membranes. Further, ALA has an important role in the production of glutathione, one of the "primary" antioxidants produced directly by the body.

Where ALA surpasses other conventional antioxidants is that it is targeted to the mitochondria. Most other antioxidants can't effectively concentrate at the level of the mitochondrion and so are almost meaningless in their ability to support the most important generator of free radicals.

ALA and Its Role in NAD Biology

Another benefit of ALA is its ability to modulate the state of the energy carrier nicotinamide adenine dinucleotide (NAD). For example, when exposed to high levels of glucose, cells are not able to properly "discharge" NADH (the electron-carrying form) to NAD^+ (its free form). The resulting imbalance of NADH to NAD^+ creates an undesirable situation in the cell.

First, the cell is denied access to the free NAD^+ (it needs this for a number of essential functions, including the proper uptake and utilization of glucose and protein for fuel).

Second, the excess NADH leads to free-radical damage through two distinct mechanisms. Excess NADH causes a breakdown of the cell's iron stores, accelerating the production of free radicals. However, even more concerning is that excess NADH, in the absence of sufficient numbers of ETCs, causes the mitochondria to become backed up with excessive electrons. Remember, NADH enters the ETC at Complex I, and this is the primary location where excess electrons are fumbled, react with oxygen, and generate superoxide radicals. ALA resolves this metabolic mess by helping to restore the balance of the two forms of NAD.

There's also another way that ALA can modulate the NADH:NAD^+ ratio, and even influence the aging process at the cellular level—through the activation of a class of genes known as *sirtuins*. These genes have been shown to be important antiaging genes in a wide range of organisms, including humans. Sirtuins are involved in the regulation of energy metabolism and longevity. In fact, the activation of sirtuins might be responsible for the broad-acting antiaging effects of calorie restriction, which is so far the only proven way to slow down biological aging in mammals. As it turns out, the availability of NAD^+ is critical to the antiaging effects of sirtuins (while excessive NADH inhibits them). If ALA can boost the cellular levels of free NAD^+ while lowering NADH, it might be able to facilitate sirtuins' antiaging activity, providing a second pathway whereby ALA could influence the aging process.

Further, some exciting animal research has shown that supplementing the diet with ALA—especially when combined with acetyl-L-carnitine—can have profound antiaging benefits by restoring youthful activity levels, cognitive performance, and heart function.

R(+) Alpha-Lipoic Acid and Its Stability

I'll add just a couple of quick notes on the types of nutritional supplements you will find on the market. The body can only use one form, what we call the R(+) form. Many commercial ALA products are synthetic, and they contain the inactive S(-) isomer in equal parts to the biologically active R(+) isomer, which means you're only getting 50 percent of the possible biological activity from the product.

Also, ALA is not stable at room temperature for long. What happens is that elevated temperatures cause ALA molecules to *polymerize* (link together to form chains), and this form doesn't seem to be absorbable. For this reason, a stabilized R(+) ALA product or one that is sold from the fridge (and stored in the fridge at home) is best. At the very least, definitely avoid exposing ALA to any sources of heat (such as leaving the bottle in your car on a hot summer day).

Creatine

Most people think of creatine simply as a supplement that bodybuilders and other athletes use to gain strength and muscle mass. While it does provide these benefits, a substantial body of evidence has found that it has a wide variety of other health benefits, too.

In fact, creatine is being studied as a nutritional therapy that might help with diseases affecting the neuromuscular system, such as muscular dystrophy, Huntington's disease, Parkinson's disease, and even ALS. Other studies suggest creatine might have therapeutic benefits for wasting syndromes and muscle atrophy in the aging population, fatigue and fibromyalgia, and some brain-related disorders.

What Is Creatine?

The human body creates creatine from the amino acids *methionine*, *glycine*, and *arginine*. On average, a person's body contains about 120 grams of creatine stored in the form of creatine phosphate (also known as phosphocreatine). Certain foods (such as beef and fish) have a relatively high creatine content.

Creatine is directly related to ATP. When a cell uses ATP, it loses a phosphate molecule and becomes ADP, which must be converted back to ATP for useful energy cycling. Because creatine is stored in the body as creatine phosphate, it can donate a phosphate molecule to ADP to regenerate ATP. This process is very speedy and is the main source of cellular energy production at the start of high-intensity anaerobic activity (such as a 100-meter sprint or lifting heavy weights). Having a large pool of creatine phosphate means this fast pathway of ATP regeneration can be sustained longer, which is exactly why creatine has been so beneficial for athletes.

A significant body of research has shown that administering supplemental creatine (usually as creatine monohydrate) can increase the total body pool of creatine phosphate. Although this increase leads to positive outcomes for energy generation and performance during explosive, anaerobic forms of exercise, the benefits of creatine in longer-duration activities and sports (such as long-distance running, rowing, and swimming) are questionable at this time. Of course, when you consider the biochemistry, creatine's benefits (and lack of benefits) in certain situations make sense.

Creatine and the Brain

Because the brain and nervous system require such high amounts of energy, it's logical to assume the neurological system can benefit greatly from creatine—and this is being corroborated by clinical research. A growing number of studies have found that creatine can protect the brain from neurotoxic agents and certain forms of brain injury.

Studies have found creatine to be highly neuroprotective against various neurotoxic agents, including MPTP (a chemical that impairs energy production in brain cells and has been used in lab animals to induce Parkinson's). Other studies found that creatine protected nerve cells from ischemia-related damage similar to what is often seen after a stroke.

Impressive? The neurological benefits don't stop there. As mentioned earlier, other studies have found that creatine can play a therapeutic or protective role in Huntington's disease and ALS. Research is promising; all this could be just the tip of the iceberg.

Creatine and Neuromuscular Diseases

One of the most promising areas of research for creatine is its effect on neuromuscular diseases such as muscular dystrophy, where researchers found a mild but significant improvement in muscle strength. Patients' everyday activities were also generally improved, and the supplemental creatine was well tolerated.

Creatine and Heart Health

Cardiac cells are also dependent on large amounts of ATP to function properly, and studies show that creatine levels are depressed in patients with heart failure. It is well known that people suffering from chronic heart failure have limited endurance and strength and tire easily, which greatly limits their ability to function in everyday life. Therefore, researchers have looked at supplemental creatine to improve heart function and overall endurance in certain forms of heart disease. While the results are not entirely conclusive at this point, clinical studies have shown increases in strength and endurance in heart failure patients.

B Vitamins

Of all the nutrients defined as true vitamins, the ones that have the greatest direct impact on cellular metabolism and energy production are collectively known as the B vitamins. This group is made up of numerous distinct nutrients, and each is either a cofactor in an important metabolic process or a precursor of an important energy-related molecule.

Vitamin B_1

Vitamin B_1 is also known as *thiamine,* and the active form is called *thiamine pyrophosphate* (TPP). TPP functions in carbohydrate metabolism to help convert pyruvate to acetyl-CoA for entry to the TCA cycle and subsequent steps to generate ATP. Because of this role, thiamine also functions in maintaining the nervous system, memory, and heart muscle health.

A deficiency of thiamine causes a condition known as *beriberi*, which is now predominantly seen only in alcoholics. Thiamine deficiency also results from excessive vomiting and often goes unrecognized until the symptoms are dramatic—and by then, might be irreversible.

The major symptoms of beriberi involve the brain and nervous system, heart, and muscles (all energy-intense organs). Brain-related effects include sensory disturbances and impaired memory. When beriberi affects the heart, symptoms are shortness of breath, palpitations, and, eventually, heart failure.

Dietary thiamine requirements are based on caloric intake; those individuals who consume more calories, such as athletes, are likely to require a higher-than-average intake of thiamine to help convert the extra carbohydrates into energy.

Vitamin B₂

Riboflavin, also known as vitamin B_2, is a major component of the cofactors FMN (also known as riboflavin-5-phosphate, in Complex I) and FAD (in Complex II). The major role of FAD in the mitochondria is to shuttle energy (electrons) from the TCA cycle and beta-oxidation to Complex II of the ETC. Because both Complex I and II pass their electrons to Complex III via CoQ10, for patients with a Complex I deficiency, riboflavin might theoretically help bypass this problem by channeling the flow of electrons through Complex II.

Vitamin B₃

Vitamin B_3 comes in different forms; *niacinamide* (also called *nicotinamide*) and *niacin* (nicotinic acid) are the predominant forms found in supplements. Based on numerous clinical trials, niacin (but not niacinamide) appears to be a relatively safe, inexpensive, and effective treatment for high LDL and low HDL cholesterol. On the other hand, niacinamide (but not niacin) has been investigated for the prevention and delay of type 1 diabetes and as a treatment for osteoarthritis.

Back in 2001, NASA scientists found traces of vitamin B_3 in meteorites, adding support to the theory that life on Earth was seeded by extraterrestrial sources. Vitamin B_3's importance to life is apparent when we consider that it is the precursor of the biological molecules NAD^+ and NADH (see "Alpha-Lipoic Acid," page 159). Without NAD^+ and NADH, mitochondria would not function in the way they do, and a major percentage of ATP would not be produced. As a precursor to NAD^+, vitamin B_3 is perhaps the single greatest nutrient for NAD biology. As the rate-limiting cosubstrate for the sirtuin enzymes, NAD modulation is emerging as a valuable tool in regulating sirtuin function and, consequently, oxidative metabolism and protection against metabolic diseases.

Recently, more biologically efficient forms of vitamin B_3 have emerged. For example, *nicotinamide riboside* currently appears as the most efficient precursor of NAD^+ and NADH. Nicotinamide riboside is found naturally in trace amounts in milk and other foods, and is a more potent version of niacin and niacinamide because it enters the biochemical pathway *after* the rate-limiting step in NAD synthesis.

Related to its role in energy metabolism and increased activity of mitochondria, interest is growing in the use of vitamin B_3 for the treatment of neuropathies and neurodegenerative diseases, diabetes, cancers, and inflammation. Other benefits include increased fatty acid oxidation, resistance to the negative consequences of high-fat diets, antioxidant protection, prevention of peripheral neuropathy, and reduction of muscle degeneration.

Vitamin B_5

Vitamin B_5 is known as *pantothenic acid* or *pantethine* (the latter being the coenzyme form). It has a very important role in the body as a precursor of coenzyme A (CoA), which is essential for the metabolism of carbohydrates, the synthesis/degradation of fats, and the synthesis of sterols (which produce steroid hormones including melatonin). It is also important for the synthesis of the neurotransmitter *acetylcholine* (important for memory) and *heme*, a component of hemoglobin (which carries oxygen—for oxidative phosphorylation—to the cells), which I will discuss in the next section, "Iron," page 167). Detoxification of many drugs and toxins requires CoA in the liver.

However, specific to cellular energy metabolism, CoA is what allows the end-product of glycolysis (pyruvate) to enter the cycle.

Therefore, as a precursor of CoA, vitamin B_5 plays a critical role in allowing energy production to occur through aerobic metabolism in the mitochondria, and not just anaerobic metabolism in the cytosol.

Vitamin B₆

Vitamin B_6 is also known as *pyridoxine*, and *pyridoxal-5-phosphate* is its active form. It's necessary for the proper function of over seventy different enzymes that participate in energy metabolism (among other things). It is also involved in the synthesis of neurotransmitters in the brain and nerve cells, and it might support mental function (mood) and nerve conduction. It might also improve emotional outlook and mood through serotonin synthesis. It is necessary for hemoglobin synthesis and red blood cell growth, which is critical for proper delivery of oxygen to the mitochondria.

Vitamin B₁₂

Vitamin B_{12}, also known as *cobalamin*, is the only vitamin containing a trace mineral: cobalt. The two metabolically active forms are *methylcobalamin* and *adenosylcobalamin* (the latter is the predominant form found in the mitochondria).

Cobalamin owes its name to the cobalt at the center of its molecular structure. Although humans require cobalt, it's only assimilated in the form of vitamin B_{12} (not free cobalt, which can be toxic). Cobalamin is found in a variety of foods, such as fish, shellfish, meat, eggs, and dairy products—but negligible amounts are found in vegetarian sources, which is why it's important for strict vegetarians and vegans to be aware of their vitamin B_{12} status. Also, the absorption of vitamin B_{12} requires another compound called *intrinsic factor* for absorption.

Vitamin B_{12} plays an important role in supplying essential methyl groups for protein and DNA synthesis, and has numerous functions. However, for the mitochondria, vitamin B_{12} is involved in several important metabolic processes, including the generation of S-*adenosyl methionine* (SAMe), which is important for cell function and survival. In turn, SAMe has a number of functions itself, but also supports the formation of creatine, the precursor of creatine phosphate, as discussed earlier. It's also a part of various protein subunits that make up the complexes in the ETC. These last two reasons are likely why many people who receive vitamin B_{12} injections often report increased levels of energy afterward.

Iron

Iron is an essential mineral and an important component of proteins involved in oxygen transport and metabolism. Heme, which is the major functional form of iron, is synthesized by the mitochondria. Heme is a critical component of hemoglobin, which is present in red blood cells and picks up oxygen as it travels through the lungs, delivering it to the cells. Heme also makes up *myoglobin*, which is similar to hemoglobin but is found in skeletal muscles. Heme is also an essential component of various proteins within the complexes of the ETC (along with cobalt, discussed in "Vitamin B_{12}"). Studies have shown that when heme metabolism is disrupted, the result is mitochondrial decay, oxidative stress, and iron accumulation, all of which are hallmarks of aging.

Biosynthesis of heme requires vitamins B_2, B_5, B_6, biotin, alpha-lipoic acid, and the minerals zinc, iron, and copper. These nutrients are essential for the production of succinyl-CoA (the precursor of heme) by the TCA cycle. Therefore, the mitochondrial pool of succinyl-CoA might limit heme biosynthesis when nutrient deficiencies exist, especially iron.

The World Health Organization considers iron deficiency to be the largest international nutritional disorder. Approximately 50 percent of anemia worldwide is attributable to iron deficiency, and it's particularly common among women of child-bearing ages.

If you're deficient in iron, not only do you have reduced capacity in the blood to deliver a critical substrate of oxidative phosphorylation —oxygen—but the functioning of your mitochondria in general could be compromised due to a potential reduction in the number of ETC complexes in operation. Iron-deficient people often report an increase in energy once their deficiency is corrected.

A word of caution—and a critical one at that: Don't supplement with iron unless you need to take iron according to a blood test.

Too much iron can increase the free-radical burden within the body and lead to a number of health conditions. Iron accumulation has been linked to neurodegenerative diseases such as Parkinson's and Alzheimer's, and can even result in death (it is not uncommon for children to overdose on iron). Just like everything else in life, moderation is key.

Resveratrol and Pterostilbene

Resveratrol, the compound made famous by the French Paradox, has been receiving significant interest from the scientific community for its antiaging benefits. While its popularity as a supplement has decreased since its peak, the scientific community continues its investigations. Scientists have also discovered another closely related compound called *pterostilbene* (pronounced "ter-ro-STILL-been"), and this is likely the "next resveratrol" that will be highly sought after at your local health food store. Pterostilbene is mainly found in blueberries, but also in grapes and the bark of the Indian Kino tree (used for centuries in Ayurvedic medicine—traditional medicine in India).

Resveratrol and pterostilbene are closely related and classified as "stilbene" compounds. Due to the similarity in their chemical structure, they have similar functions, but they are not identical. What's interesting, however, is that these two compounds work in a synergistic fashion. Pterostilbene produces its beneficial effects on gene expression in ways that enhance those produced by resveratrol.

One of the major benefits of resveratrol and pterostilbene is the ability to mimic many of the beneficial effects of calorie restriction (discussed in the following section in detail) by favorably regulating genes involved in the development of cancers, atherosclerosis, diabetes, and the systemwide inflammation that underlies a variety of age-related disorders.

Research has found that resveratrol activates genes near the start of the molecular cascade precipitated by caloric restriction, while pterostilbene directly activates genes downstream from resveratrol's action. This synergistic and complementary action might help prevent

Turning On or Off Our Genetic Light Switches

Promising research into resveratrol and pterostilbene has revealed that their benefits are not due to their being antioxidants (although they seem to have this effect in a test tube), but instead are due to their ability to turn "on" or "off" certain genes. Many people who learned about genetics decades ago still view genes as *fixed* units of information that are inherited from our parents and determine physical characteristics such as eye color. As our understanding of genetics expands rapidly, what's apparent is that we can actually *modify* the meaning behind those genes. This process is known as *gene expression*, the study of which is called *epigenetics*. The process occurs when stimuli originating from within or outside our bodies (e.g., diet, environmental toxins, or endogenous factors such as stress, sleep patterns) switch certain genes on or off. In epigenetics, the goal of manipulating various nutrients and lifestyle factors is to switch on protective genes and switch off harmful ones. This is one of the most exciting areas of medical research today.

cancers and diabetes, support healthy blood lipids, and produce longevity-promoting effects across the cycle of gene expression.

Whether we're talking about the fat-sensing complexes found to favorably affect lipid profiles, modifying several vital glucose-regulating enzymes (which help to control blood sugar), reducing the production of inflammatory mediators, or up-regulating specific brain proteins associated with improved memory, pterostilbene produces beneficial changes almost identical to those seen in calorie restriction. Of course, many of these benefits relate back to the humble mitochondria.

Ketogenic Diets and Calorie Restriction

Ketone bodies, herein also referred to simply as *ketones*, are three water-soluble compounds that are produced as by-products when fatty acids are broken down for energy in the liver. These ketones can be used as a source of energy themselves, especially in the heart and brain, where they are a vital source of energy during periods of fasting.

The three endogenous ketones produced by the body are *acetone*, *acetoacetic acid*, and *beta-hydroxybutyric acid* (which is the only one that's not technically a ketone, chemically speaking). They can be converted to acetyl-CoA, which then enters the TCA cycle to produce energy.

Fatty acids are so dense in energy, and the heart is one of the most energy-intensive organs, so under normal physiologic conditions, it preferentially uses fatty acids as its fuel source. However, under ketotic conditions, the heart can effectively utilize ketone bodies for energy.

The brain is also extremely energy-intensive, and usually relies on glucose for its energy. However, when glucose is in short supply, it gets a portion of its energy from ketone bodies (e.g., during fasting, strenuous exercise, low-carbohydrate, ketogenic diet, and in neonates). While most other tissues have alternate fuel sources (besides ketone bodies) when blood glucose is low, the brain does not. For the brain, this is when ketones become essential. After three days of low blood glucose, the brain gets 25 percent of its energy from ketone bodies. After about four days, this jumps to 70 percent!

In normal healthy individuals, there is a constant production of ketone bodies by the liver and utilization by other tissues. Their excretion in urine is normally very low and undetectable by routine urine tests. However, as blood glucose falls, the synthesis of ketones increases, and when it exceeds the rate of utilization, their blood concentration increases, followed by increased excretion in urine. This state is commonly referred to as *ketosis*, and the sweet, fruity smell of acetone in the breath is a common feature of ketosis.

Historically, this sweet smell was linked to diabetes and ketones were first discovered in the urine of diabetic patients in the mid-nineteenth century. For almost fifty years thereafter, they were thought to be abnormal and undesirable by-products of incomplete fat oxidation.

In the early twentieth century, however, they were recognized as normal circulating metabolites produced by the liver and readily utilized by the body's tissues. In the 1920s, a drastic "hyperketogenic" diet was found to be remarkably effective for treating drug-resistant epilepsy in children. In 1967, circulating ketones were discovered to replace glucose as the brain's major fuel during prolonged fasting. Until then, the adult human brain was thought to be entirely dependent upon glucose.

During the 1990s, diet-induced *hyperketonemia* (commonly called *nutritional ketosis*) was found to be therapeutically effective for treating several rare genetic disorders involving impaired glucose utilization by nerve cells. Now, growing evidence suggests that mitochondrial dysfunction and reduced bioenergetic efficiency occur in brains of patients with Parkinson's disease and Alzheimer's disease. Since ketones are efficiently used by brain mitochondria for ATP generation and might also help protect vulnerable neurons from free-radical damage, ketogenic diets are being evaluated for their ability to benefit patients with Parkinson's and Alzheimer's diseases, and various other neurodegenerative disorders (with some cases reporting remarkable success).

There are various ways to induce ketosis, some easier than others. The best way is to use one of the various ketogenic diets (e.g., classic, modified Atkins, MCT or coconut oil, low-glycemic index diet), but calorie restriction is also proving its ability to achieve the same end results when carbohydrates are limited.

Features of Caloric Restriction

There are a number of important pieces to caloric restriction. First, and the most obvious, is that caloric intake is most critical. Typically, calories are restricted to about 40 percent of what a person would consume if food intake was unrestricted. For mice and rats, calorie restriction to this degree results in very different physical characteristics (size and body composition) than those of their control-fed counterparts. Regarding life extension, even smaller levels of caloric restriction (a reduction of only 10–20 percent of unrestricted calorie intake) produce longer-lived animals and disease-prevention effects.

In April of 2014, a twenty-five-year longitudinal study on rhesus monkeys showed positive results. The benefit of this study was that it was a long-term study done in primates—human's closest relatives—and confirms positive data we previously saw from yeasts, insects, and rodents. The research team reported that monkeys in the control group (allowed to eat as much as they wanted) had a 2.9-fold increased risk of disease (e.g., diabetes) and a 3-fold increased risk of premature death, compared to calorie-restricted monkeys (that consumed a diet with 30 percent less calories).

If other data from studies on yeast, insects, and rodents can be confirmed in primates, it would indicate that calorie restriction could extend life span by up to 60 percent, making a human life span of 130–150 years a real possibility without fancy technology or supplements or medications. The clear inverse relationship between energy intake and longevity links its mechanism to mitochondria—energy metabolism and free-radical production.

Second, simply restricting the intake of fat, protein, or carbohydrates without overall calorie reduction does not increase the maximum life span of rodents. It's the calories that count, not necessarily the type of calories (with the exception of those trying to reach ketosis, where type of calorie does count).

Third, calorie restriction has been shown to be effective in disease prevention and longevity in diverse species. Although most caloric restriction studies have been conducted on small mammals like rats or mice, caloric restriction also extends life span in single-celled protozoans, water fleas, fruit flies, spiders, and fish. It's the only method of life extension that consistently achieves similar results across various species.

Fourth, these calorie-restricted animals stay "biologically younger" longer. Experimental mice and rats extended their youth and delayed (even prevented) most major diseases (e.g., cancers, cardiovascular diseases). About 90 percent of the age-related illnesses studied remained in a "younger" state for a longer period in calorie-restricted animals. Calorie restriction also greatly delayed cancers (including breast, colon, prostate, lymphoma), renal diseases, diabetes, hypertension, hyperlipidemia, lupus, and autoimmune hemolytic anemia, and a number of others.

Fifth, calorie restriction does not need to be started in early age to reap its benefits. Initiating it in middle-aged animals also slowed aging (this is good news for humans, because middle age is when most of us begin to think about our own health and longevity).

Of course, the benefits of calorie restriction relate back to mitochondria. Fewer calories mean less "fuel" (as electrons) entering the ETC, and a corresponding reduction in free radicals. As you know by now, that's a good thing.

Health Benefits

As just discussed, new research is showing that judicious calorie restriction and ketogenic diets (while preserving optimal nutritional intake) might slow down the normal aging process and, in turn, boost cardiovascular, brain, and cellular health. But how? We can theorize that the restriction results in fewer free radicals, but one step in confirming a theory is finding its mechanism.

In particular, researchers have identified the beneficial role of *beta-hydroxybutyric acid* (the one ketone body that's not actually a ketone). It is produced by a low-calorie diet and might be the key to the reduced risk of age-related diseases seen with calorie restriction. Over the years, studies have found that restricting calories slows aging and increases longevity, but the mechanism behind this remained elusive. New studies are showing that beta-hydroxybutyric acid can block a class of enzymes, called *histone deacetylases*, which would otherwise promote free-radical damage.

While additional studies need to be conducted, it is known that those following calorie-restricted or ketogenic diets have lower blood pressure, heart rate, and glucose levels than the general population. More recently, there has been a lot of excitement around *intermittent fasting* as an abbreviated method of achieving the same end results.

However, self-prescribing a calorie-restricted or ketogenic diet is not recommended unless you've done a lot of research on the topic and know what to do. If not done properly, these diets can potentially increase mental and physical stress on the body. Health status should be improving, not declining, as a result of these types of diets, and when not done properly, these diets could lead to malnutrition and starvation. Health care practitioners also need to properly differentiate a patient who is in a deficiency state of anorexia or bulimia versus someone in a healthy state of ketosis or caloric restriction.

I'll add a final word of caution: While ketogenic diets can be indispensable tools in treating certain diseases, their use in the presence of mitochondrial disease—at this point—is controversial and depends on the individual's specific mitochondrial disease. In some cases, a ketogenic diet can help; in others it can be deleterious. So, of all the therapies listed in this book, the one for which I recommend specific expertise in its application is this diet, and only after a proper diagnosis.

Diabetes

As previously discussed, diabetes and insulin resistance are huge and growing problems in our society, and new research on them points back to the mitochondria. As we age, our mitochondria deteriorate, which means that the rates of fat oxidation and subsequent production of energy also slow down. This slowdown predisposes us to fat accumulation in muscle and possibly in the liver—two organs sensitive to insulin. It also means the mitochondria in the beta cells (the cells in the pancreas responsible for insulin production and secretion) slow down and diminish in health over time. Beta-cell dysfunction then results in impaired glucose tolerance, and then, ultimately, type 2 diabetes.

The great news is that this downward spiral can be stopped—perhaps even reversed—by a combination of calorie restriction and exercise. A study published in 2007 found that middle-aged obese diabetics who cut their calorie intake by 25 percent and did moderate-intensity exercise (like walking) on most days for four months boosted their mitochondrial density by 67 percent, which ultimately improved their insulin sensitivity by 59 percent.

Another pilot study published in 2011 showed that an extreme calorie-restricted diet (only 600 kilocalories per day) for eight weeks *reversed* diabetes in 100 percent of the study's participants. This pilot study has single-handedly altered our understanding of type 2

Ketones and Cancer

Exciting research also continues into the use of a ketogenic diet as a drug-free cancer treatment (or an adjunct to conventional treatment). The diet calls for eliminating carbohydrates and replacing them with healthy fats and proteins (a modified Atkins diet). The premise is that because cancer cells need glucose and carbohydrates to thrive, eliminating these two things literally starves the cancer cells.

Calorie restriction also works to reduce cancer risk in the first place. Though there are other ways calorie restriction works to improve mitochondrial and cellular health (such as by activating sirtuin 1, blocking histone deacetylases), the simplest way to explain how it reduces cancer risk is that less calories consumed means less electrons entering the ETC.

This area is seeing a significant amount of scientific and clinical attention, and the few pages I dedicated to these two diets obviously don't do it justice, and thus, I encourage you to learn more on your own. Most people eat more calories daily than physiologically necessary, so dropping some calories from the diet should be a real consideration for most. While many scientists have believed for many years that adopting a low-calorie diet—while still maintaining proper nutrition—could delay the effects of aging, they've only recently found out why. Expect more studies to be published focused on this topic.

Be mindful of your standard diet and consider cutting things back slowly, while continuing to ensure all your nutritional needs are met. It's never too late to start, although the earlier the better! Besides, I think everyone would welcome a 30 percent savings off their grocery bill, and side benefits to the planet also include better food security and less need for genetically modified organisms (if you believe Big Agri's propaganda that GMOs are the answer to feeding a ballooning global population).

For those who are interested in more information, I'd suggest the books *The Alzheimer's Antidote: Using a Low-Carb, High-Fat Diet to Fight Alzheimer's Disease, Memory Loss, and Cognitive Decline* by Amy Berger (Chelsea Green, 2017) and *Keto for Cancer: Ketogenic Metabolic Therapy as a Targeted Nutritional Strategy* by Miriam Kalamian (Chelsea Green, 2017).

diabetes, which was thought to be a lifelong, irreversible condition. What's more is that 64 percent of the subjects in the study remained diabetes-free three months after the end of dietary intervention. Perhaps as a consequence of the reduction in fat content, the pancreas was found to regain its normal ability to produce insulin, and muscles and liver once again became sensitive to insulin. As a result, blood sugar levels after meals began to improve steadily.

Additionally, a study out of Hamilton, Ontario (Canada), published in March 2017, confirmed these results. This study, involving eighty-three diabetic participants, found that just four months of "intensive metabolic intervention" (calorie restriction and exercise) not only resulted in amazing results to blood sugar and body weight during the study's intervention period, but over 40 percent of participants continued to be diabetes-free twelve weeks after the end of the metabolic intervention.

Given these results in a growing number of studies, it appears that soon we might need to reclassify diabetes as a metabolic disease rather than an endocrine disorder—it's not insulin resistance that's the root of the problem, but poor mitochondrial health and metabolism. This gives immeasurable hope to those with the condition. Of course, the results of these "pilot studies" need to be replicated in much larger clinical trials, but the new data we're seeing is extremely exciting and potentially paradigm shifting.

Massage and Hydrotherapy

An interesting study from Hamilton, Ontario (Canada), found that therapeutic massage can increase the biomarkers of mitochondrial biogenesis. In addition, it was previously thought impossible to increase brown adipose tissue (BAT), because we assumed only newborns and young children had it. This has changed as we recently discovered that BAT is present and active in adults, where it is situated predominantly around the aorta and in the supraclavicular area (neck area). BAT volume and activity are lower in individuals who are obese, which might confirm the theory that BAT significantly contributes to total energy expenditure.

What's more, however, is that various ways in which BAT can be manipulated to increase the expenditure of energy have been identified. It has been shown that white adipose tissue (WAT) can undergo a process known as *browning* where it takes on characteristics of BAT. This happens with physiological or biochemical stimulation (such as chronic cold exposure, hormonal stimuli, or pharmacological treatment). These "inducible brown fat cells" (also known as *beige fat*) normally have low thermogenesis activity and a low number of mitochondria; however, once activated, they possess many biochemical and morphological features of BAT, such as the presence of multilocular lipid droplets, an abundance of mitochondria (mitochondrial biogenesis), and higher levels of UCP1.

Methods on how to achieve this include repeated and/or chronic exposure to cold temperatures (such as being out in the cold more during winter months, turning down the thermostat in your home in the winter, or a number of other methods, such as are traditionally done in hydrotherapy). A study published in the summer of 2014 found that sleeping with the temperature set at 19°C (versus 24°C) induced BAT in an adult population, and consequently improved insulin sensitivity and glucose metabolism. What an easy way to save on your winter heating bills and simultaneously do something great for your health!

Going from hot sauna to cold shower repeatedly could increase BAT over time. Similarly, just finishing your shower with a cold rinse could do the same. However, neither of these suggestions has been specifically proven in a clinical setting, and I think more research needs to be done to confirm hydrotherapy's effects on mitochondria and BAT or UCP1 production, but it does seem encouraging. In fact, I expect lots of research to be done in this area in the coming years as inducing BAT is a major target in anti-obesity research.

Cannabis and Phytocannabinoids

I believe science and research are finally shifting people's opinion about cannabis, and with legalization efforts in Canada and a number of US states, we're seeing a much greater acceptance of this botanical. After many decades of prohibition-style propaganda, the truth is bubbling to the top, and I expect a significant amount of research on the health benefits of cannabis to come to light. For mitochondrial health, the research on cannabis is still in its early stages, but it shows great promise.

Present in all mammals, the endocannabinoid system (ECS) is considered the master regulator of homeostasis in the body, and we all produce our own cannabinoid compounds (called *endocannabinoids*). These compounds work in the same way as the *phytocannabinoids* found in Nature, such as those in the cannabis plant. When our body can't produce enough endocannabinoids to meet its needs, phytocannabinoids become increasingly important.

In 2012, French scientists discovered that mitochondria contained cannabinoid receptors on their membranes. This revelation laid the groundwork for successive investigations into the role of the ECS in regulating mitochondrial activity. Since the ability to conduct research into this incredible plant has become easier, scientific evidence is showing that *cannabidiol* (CBD) and *tetrahydrocannabinol* (THC)—the two main phytocannabinoids from the cannabis plant—can directly and indirectly impact the mitochondria. It turns out that many of the biological processes that involve mitochondria are modulated by endo- and phytocannabinoids.

On the surface, research on cannabinoids often shows contradictory results. Cannabinoids are notorious for exerting opposite effects under different situations. CBD and THC have been shown to balance physiological excesses as well as deficiencies. A small dose of cannabis stimulates while a large dose tends to sedate. Phytocannabinoids can destroy cancer cells but leave healthy cells alone. While these aspects might be expected by those who know the "modulating" or balancing effects of cannabinoids, newer research is revealing how the mitochondria can answer these and other confusing aspects of the ECS.

Cannabinoids are known to promote homeostasis through various bidirectional pathways. While most drugs have a fairly linear dose-response curve (a higher dose causes a stronger effect), cannabinoids have a biphasic dose-response curve. A biphasic effect means two opposite responses are possible from a single compound (this is not uncommon, especially among cannabinoids). For example, low doses of THC tend to increase mitochondrial activity, while higher doses tend to decrease mitochondrial activity. Typically, these biphasic effects of cannabinoids depend on the strength of the signal (dose), as well as the context in which the signal is overlaid (what else is going on in the mitochondria and cell).

This might seem reminiscent of the free-radical signal discussion earlier in the book (see "A Radical Signal," page 34). In fact, growing evidence suggests that cross-talk between the ECS and the free-radical signaling systems acts to modulate functionality of both the ECS and redox homeostasis. Further, as just discussed, studies reveal that interactions between the ECS and free-radical signaling systems can be both stimulatory and inhibitory, depending on cell stimulus, the source of free radicals, and cell context. While such cross-talk might act to maintain cell function, abnormalities in either system could propagate and undermine the stability of both systems, thereby contributing to various pathologies associated with their dysregulation.

Other research suggests that THC can inhibit the formation of amyloid plaques in the brain by enhancing mitochondrial function. Further, CBD has been shown to induce mitochondrial biogenesis and reverse memory loss in animals.

Investigations into CBD's effect on mitochondria have shed light on how it can protect against brain injury by regulating fluctuations in intracellular calcium. These results could be good news for future stroke victims and might suggest that CBD could reduce the severity of ischemic damage (by modulating intracellular calcium ions). Further, in 2017, a study found that an imbalance of calcium ions in the mitochondria might drive Alzheimer's disease, further strengthening the connection between the benefits of cannabis and this debilitating disease.

The preceding discussion is just a glimpse into some of the research on the two most prominent cannabinoids, yet there are many more cannabinoids found in cannabis, as well as many *terpenes* (another large class of compounds that give cannabis its therapeutic benefits). In reality, as with all botanical medicines, the whole is greater than the sum of its parts, so I'm excited to see more sophisticated research on the entire plant with its full spectrum of beneficial compounds, including terpenes (instead of just isolated compounds such as THC or CBD).

Epidemiological studies can help—and possibly hint at what's going on. For example, a study published in 2013 confirmed cannabis's benefits to cardiovascular health, insulin levels, HDL cholesterol, and waist circumference—adding significance to previous epidemiological studies that found lower rates of obesity and diabetes in cannabis users. When we consider the importance of mitochondrial function in these health conditions, and with the new research showing the intricate relationship between the ECS and mitochondria, it's time we shed the stigma around cannabis and embrace it for its health-supportive properties.

Exercise and Physical Activity

Exercise and physical activity are the final topic I'll cover, but it is probably the most important when it comes to mitochondrial health. There is an intriguing paradox here, however. In recent years, strenuous physical sports such as the ultramarathon, cross-country running, and iron man triathlon have become increasingly popular around the world. We've been beaten to death over recent years with the "exercise is healthy" message, and together with our natural predisposition to assume more of something healthy is better, the increasing popularity of these sports should be welcome news. Unfortunately, that's not the case.

Strenuous or exhaustive exercise-related muscle damage has been associated with a high degree of oxygen consumption and free-radical damage and an increase in the pro-inflammatory mediators as indicated by muscle soreness, swelling, prolonged loss of muscle function, and leakage of muscle proteins and *nucleotides* (the energy pool) into circulation, among other effects. In addition to the muscles, during exercise, many other tissues can produce high amounts of free radicals, such as the heart, the lungs, and even blood.

So if strenuous, exhaustive exercise is not good, and being sedentary is no better, it's sensible that a moderate amount of nonexhaustive exercise is the ticket.

Just as moderation is the key to many—if not all—things deemed healthy, exercise is no different. The beneficial effects of exercise are lost with strenuous or exhaustive exercise. And if inadequate time is given for recovery, things get even worse (some scenarios specific to loss of the energy pool were discussed in "D-Ribose," page 132).

The beneficial effects of regular, nonexhaustive physical activity have been known for a long time. Regular exercise is associated with diverse health benefits, such as reduced threat of cardiovascular diseases, cancers, diabetes, and, in general, a lower risk of all-cause mortality. An interesting study published in 2014 also showed exercise can lower the risk of age-related macular degeneration; a separate study, also published in 2014, showed that being sedentary ranks higher than smoking, obesity, or high blood pressure as a risk factor for heart disease. What this latter study would suggest is that a physically active smoker is actually healthier than a sedentary nonsmoker! Yes, exercise is *that* important.

Yet the problem still remains; even moderate-intensity, nonexhaustive exercise increases the rate at which free radicals are produced in the mitochondria. So, at least on the surface, all physical activity should be minimized. Unfortunately, no such luck for the couch potatoes. The fact that exercise promotes health and longevity is what's been called the Exercise Paradox.

It is a paradox because exercise induces significant amounts of free-radical production. While this free-radical production should have negative effects on the health of mitochondria, it does not. Moderate levels of physical activity result in moderate levels of free radicals. Not only does exercise increase the energy demand, which results in mitochondrial fission and biogenesis (through signaling from the relative abundance of AMP versus ATP, and various other mechanisms, such as increased expression of PGC-1alpha and PPAR-gamma), but the free radicals that are produced send a signal to the cell indicating it needs to produce more complexes for the ETC. As I discussed near the beginning of this book, sometimes free radicals play a key role in cell signaling (see "A Radical Signal," page 34).

In response to this oxidation, the cell realizes that it needs more mitochondria—and more ETC within each mitochondrion—to be able to meet the energy demands placed on it. After repeated bouts of moderate-intensity exercise, the number of mitochondria per cell has increased, and each mitochondrion has a higher number of ETCs.

The end result of this process is that, at rest (which makes up the vast bulk of our modern day), we have an abundant amount of spare capacity in each mitochondrion and in each cell—just like birds! And so, at rest, those who are physically fit and active produce far fewer free radicals in the mitochondria than sedentary individuals do. During physical activity, physically fit and active people also produce more energy (translated as improved physical performance) while producing much fewer free radicals. It's a real-life cellular example of needing to take a step back (oxidation) to take two steps forward (improved mitochondrial function and capacity).

And so we have found the best way to be more like birds (see "How Do the Birds Do It?" page 127). Flight requires incredible amounts of energy, and over time, birds have developed the bioenergetic capacity to generate that enormous amount of energy. At rest, they have a vast amount of spare capacity, and because of this, they produce far, far fewer free radicals during the majority of their lives.

The other benefit is that all this physical activity also uses up the ATP. If we don't use up the ATP, we end up with a backlog of energy, and the electrons in ETC will overflow and create free radicals. However, unlike a situation where free radicals are produced in the presence of abundant ATP that's not being used—which has no benefit in stimulating

Cognitive Health and the New Frontier for Physical Activity

While much research has been conducted over the years on the cardiovascular benefits of exercise (so much in fact that I've decided against discussing it here), recent years have seen a focus on brain and cognitive function take center stage with an aging population. Both resistance and aerobic training have been shown to improve different types of memory, executive functioning, and functional plasticity.

It's been known for decades that aerobic exercise can increase the number of mitochondria in your muscle cells by up to 50 percent in as little as six weeks. To get the benefit, however, you need to do aerobic exercises (such as running, cycling, swimming, or walking briskly) at an intensity that's at least half of your maximum capacity. This intensity needs to be sustained for at least fifteen to twenty minutes per session, three to four times a week.

One of the most powerful studies on this was published in 2011, where the results showed that aerobic exercise can modify the gene responsible for producing *brain-derived neurotrophic factor* (BDNF). The researchers looked at 120 nondemented elderly individuals over a one-year period who either stretched or did aerobics. Three variables were measured: serum BDNF levels, memory function, and an analysis of the size of the hippocampus (part of the brain important for memory, and one of the first regions in the brain suffering damage in Alzheimer's). After one year, the group that did the aerobic exercise had an increase in hippocampus size by about 1 percent, improvement of memory function, and higher levels of serum BDNF! There is no pharmaceutical that can come even close to providing that much improvement.

Just plain old aerobic exercise improved memory, regenerated the hippocampus, and raised BDNF levels—which, beyond stimulating nerve growth, also stimulates *neuroplasticity* (fundamental for learning new things and adapting to stress). In 2013, the Obama administration dedicated $33 million to help pharmaceutical companies develop a drug for Alzheimer's disease prevention. Little did they know, that drug already exists, and it's called aerobic exercise! Use that $33 million and buy every adult a nice pair of running shoes and send them on a run around their neighborhood—a better spend of tax dollars, with an unbelievably high return on investment! It reminds me of a comic I read where the doctor handed a patient a pill and said, "To prevent a heart attack, take one pill every day, and take it out for a run, then take it to the gym, then take it for a bike ride . . ."

In an interesting recent study of rats, researchers even found that pregnant rats who swam improved the mitochondrial biogenesis in the brain of their offspring! It's reasonable to expect this effect would be seen in humans as well, which could have profound benefits in protecting our children from future cognitive decline and cerebral damage.

On the other hand (at least currently), the evidence is inconclusive with respect to the mitochondrial benefits of resistance exercises (such as lifting weights). In younger individuals, resistance exercises don't seem to result in increased numbers of mitochondria, but a higher mitochondrial count might be seen in older people according to at least one study.

Regardless, resistance exercises have many other benefits (such as preventing *sarcopenia*—or age-related muscle wasting—which is a big concern in older age); however, the point here is to improve cellular bioenergetics and optimize the mitochondria. For this reason, aerobic exercise is a must, with resistance exercise highly recommended the older we get.

High-intensity interval training (HIIT) is an exciting area of research. HIIT is characterized by repeated bursts of intense exercise alternating with brief periods of recovery. Sports such as hockey, lacrosse, and soccer are great examples of the HIIT style. HIIT has been shown to be much better and efficient in increasing muscle mitochondrial production and endurance per given volume of exercise (when compared to typical aerobic exercises).

mitochondrial biogenesis—free radicals produced in combination with insufficient ATP *does* trigger mitochondrial biogenesis.

So ensuring regular physical activity addresses both sides of the coin—increasing the number of mitochondria to create spare capacity, and using up ATP to prevent a backlog and overflow of electrons in the ETC.

The resulting mitochondrial biogenesis is the reason exercise has been linked to improvements in cardiovascular health, cognitive health, psychological well-being, lower diabetes risk, healthy muscles and bones, cancer prevention, reduced risk of premature death from all-cause mortality, and longevity. "Movement is medicine" is a catch-phrase we should all live by.

Pulling It All Together

So you can see there are many options readily available to most individuals who want to improve the health of their mitochondria. Not any one therapy is ideal, however, and it seems that the best outcomes are from a combination of numerous therapeutic agents and exercise (exercise must always be included in any program targeting mitochondria), especially high-intensity interval training. Other areas getting some current research attention include intermittent hypoxia and intermittent fasting—we can expect some eye-opening results to come from these studies (and their recommendations are sure to make their way into my personal mitochondrial regimen).

There are many ways to approach developing a mitochondrial regimen, depending on what health objectives you're trying to achieve and what medical history or preexisting conditions are factors. Even then, as research shapes my approach, my recommendations and personal mitochondrial regimen could change, and likely will have changed by the time you read this.

Mitochondrial medicine is a constantly evolving body of knowledge, and we're learning more about the mitochondrial benefits of different nutrients and botanicals every day. For example, a really interesting botanical I'm looking into at the moment, *Gynostemma pentaphylum*, seems to have powerful benefits for mitochondria by activating AMPK. *G. pentaphylum* has been shown to enhance mitochondrial biogenesis, reduce body fat and blood sugar, and modulate inflammation.

The field of mitochondrial medicine is exploding, and around the time of writing this book, there seemed to be between two hundred and three hundred related studies published *every week*—as there were consistently over the last five to six years I've been researching this book!

The trouble is, the more we investigate, the more complex the field gets. More recent studies are looking into the interaction of mitochondria with other cellular organelles, such as the peroxisomes and endoplasmic reticulum—both of which play a part in the health of mitochondria. In biology, there is never just one thing. For every rule that we have, we find a bunch of exceptions, and exceptions to the exceptions. As we dig deeper, we reveal a picture of complexity, rather than simplicity.

Mitochondrial medicine is a fascinating field that has real-life implications for countless health conditions, and indeed for life itself. With greater knowledge in the coming years, we might be able to nurture the Force to the level of Anakin or Luke Skywalker. Just make sure you use your newfound power for good, peace, and love. Resist the Dark Side.

ACKNOWLEDGMENTS

I'd like to give a huge thank you and my sincere gratitude to Erin, my life partner, whose support is what made this book possible. For not only holding down the home base and caring for our two sons—which allowed me to have the time to research and write—but also for being my level-headed sounding board, personal life coach, on-demand graphic designer, professional chef, and emotional bedrock whom I can always turn to.

Also would like to thank my two sons, Aidan and Hudson, for helping me grow and understand myself better. I'm constantly amazed by the empathy, integrity, and affection displayed by these two little incredible humans.

Makenna Goodman, Patricia Stone, Deborah Heimann, Nanette Bendyna, Linda Hallinger, Sean Maher, Christina Butt, and everyone at Chelsea Green Publishing for guiding me through this publishing process and helping to make the manuscript shine, and bringing its message of health to the broadest audience possible—something that would be impossible with my own limited resources.

I'd also like to acknowledge the works and publications by Nick Lane, Stephen Sinatra, Anthony Linnane, and the countless researchers, scientists, authors, and health care practitioners who have provided the knowledge, understanding, and wisdom that makes a science-heavy book like this conceivable. Scientific knowledge and understanding only progress when you get to stand on the shoulders of those who came before you.

GLOSSARY

Adenosine: a compound formed by combining a purine ring (adenine) with D-ribose.

Adenylate kinase reaction: also called myokinase reaction; the reaction where two molecules of ADP are joined to form ATP and AMP.

ADP: adenosine diphosphate, the precursor of ATP.

Aerobic metabolism: energy production in the cell through a process that uses oxygen; usually referred to as oxidative phosphorylation, which is the process of using the electron transport chain pathway in the inner mitochondrial membrane.

AMP: adenosine monophosphate; a by-product formed when two ADP combine to form ATP in the adenylate kinase reaction.

Anaerobic metabolism: energy production in the cell (mostly the cytosol) without the use of oxygen; important in providing short bursts of energy quickly, but an inefficient use of fuel that can't sustain the cell long-term.

Antioxidant: any compound that protects against oxidation (free-radical damage), either by directly sacrificing itself (to protect other molecules) or indirectly catalyzing the breakdown of biological oxidants.

Apoptosis: programmed cell death, or cellular suicide; a finely coordinated and carefully controlled mechanism for removing damaged or unnecessary cells from a multicellular organism.

Asexual reproduction: replication/propagation of a cell or organism, in which an exact duplicate of the parent cell or organism is produced.

ATP: adenosine triphosphate; the universal energy currency of life that is formed from ADP (adenosine diphosphate) and phosphate; splitting ATP releases energy used to power many different types of biochemical work, from muscular contraction to protein synthesis.

ATPase: also known as ATP synthase, this enzymatic motor is embedded within the inner mitochondrial membrane, and forms ATP (from ADP and phosphate) as protons flow through it.

Autosomal dominant: one of several ways that a trait or disorder can be inherited. If a trait or disease is autosomal dominant, an individual only needs the defective gene from one parent in order to inherit the disease.

Autosomal recessive: one of several ways that a trait or disorder can be inherited. Two copies of an abnormal gene (one from each parent) must be present in order for the disease or trait to be inherited.

Cell: the smallest biological unit capable of independent life, by means of self-replication and metabolism.

Cell wall: the tough but permeable outer "shell" of bacteria; maintains cell shape and integrity to protect from changes in physical conditions.

Chromosome: the long molecule of DNA; can be circular, as in bacteria and mitochondria, or straight, as in the nucleus of eukaryotic cells (where it is wrapped in proteins such as histones).

Cytochrome c: a mitochondrial protein that shuttles electrons from Complex III to Complex IV of the electron transport chain; when released from the inner mitochondrial membrane, cytochrome c is an important initiator of apoptosis.

Cytoplasm: everything inside the cell contained within the cell membrane, excluding the nucleus.

Cytoskeleton: a network of fibers within the cell that provides structural support; can change shape to enable some cells to move around and engulf other cells or particles.

Cytosol: the aqueous part of the cytoplasm excluding organelles, such as mitochondria, and membrane systems.

DNA: deoxyribonucleic acid, the double helix structure responsible for genes, in which nucleotide letters are paired with each other to form a template from which an exact copy of the whole molecule can be regenerated; the sequence of nucleotide letters in a gene encodes the sequence of ammo acids in a protein.

Electron: a tiny, negatively charged wave particle.

Enzyme: a protein molecule with enormous specificity, which serves as a catalyst responsible for significantly speeding up biochemical reactions.

Eukaryotic cell: cells with a true nucleus.

Free radical: a highly reactive atom or molecule with an unpaired electron.

Free-radical leakage: continuous low-level production of free radicals from the electron transport chains of the mitochondria; a result of electrons reacting directly with oxygen.

Gene: a stretch of DNA, whose sequence of letters encodes for a single protein.

Genome: the complete library of genes in an organism; the term is also taken to include noncoding (i.e., nongenic) stretches of DNA.

Heteroplasmy: a mixture of two or more different mitochondrial DNA, such as the father and the mother.

Histones: protective proteins that bind DNA in a very particular way, found mainly in eukaryotic cells.

Hypoxia: a situation that occurs when cells or tissues are deprived of oxygen.

Ischemia: a situation that occurs when there is reduced blood flow to a tissue or organ; reduced blood flow results in hypoxia.

Lateral gene transfer: the random transfer of DNA segments or genes, from one cell to another, as opposed to vertical inheritance from mother to daughter.

Lipid: a type of long-chain fatty molecule found in biological membranes and as stored fuel.

Maternal inheritance: a non-Mendelian form of inheritance in which the genotypes from one parental type are transmitted to all progeny. That is, all the genes in offspring will originate from only the mother. This phenomenon is most commonly observed in eukaryotic organelles such as mitochondria.

Membrane: the thin fatty (lipid) layer that envelops cells and forms complex systems inside eukaryotic cells.

Metabolic rate: the rate of fuel consumption or energy production, measured by the rate of glucose oxidation or oxygen consumption.

Mitochondrial DNA: the chromosome found in mitochondria; five to ten copies are typically found in each mitochondria; circular and bacterial in nature.

Mitochondrial Eve: the last female ancestor common to all humans living today, based on mitochondrial DNA that is inherited asexually down the maternal line.

Mitochondrial genes: genes encoded by mitochondrial DNA; in humans there are thirteen protein-coding genes, in addition to genes encoding the RNA ribosomes.

Mutation: an inherited or acquired change in the DNA sequence; can have a negative, positive, or neutral effect on function.

Mutation rate: the number of DNA mutations per unit time.

NADH: nicotinamide adenine dinucleotide; a molecule that ultimately carries the electrons derived from food/fuel to Complex I of the electron transport chain.

Natural selection: inheritable differences in biological adaptability to environmental conditions; the variance in individual survival and reproduction within a population.

Noncoding (junk) DNA: DNA sequences that do not code for proteins or RNA.

Nucleus: spherical membrane-enclosed "control center" of eukaryotic cells; contains chromosomes composed of DNA and protein.

Oocyte: an egg cell; the female sex cell.

Organelles: tiny organs within cells dedicated to specific tasks, such as mitochondria.

Oxidation: the loss of electrons from an atom or molecule.

Phagocytosis: the physical engulfment (by means of changing shape) of dead cells, pathogens, or particles by a cell; the particles are digested in a vacuole inside the cell.

Prokaryote: a broad class of single-celled organisms that do not possess a nucleus; includes bacteria.

Proton: the nucleus of a hydrogen atom, with a single positive charge.

Proton gradient: a difference in proton concentration between one side of a membrane and the other.

Proton pumping: the physical translocation of protons from one side of a membrane to the other.

Purines: one of the key building blocks for DNA, RNA, and ATP. Adenine is a purine.

Recombination: the physical replacement of a gene from one source with the equivalent gene from another source; takes place in lateral gene transfer, sexual reproduction, and during the repair of a damaged chromosome by reference to a spare copy.

Redox reaction: a reaction between two molecules where one is oxidized (loses an electron) while the other is reduced (gains an electron).

Redox signaling: the change in activity (usually by free radicals) of transcription factors as a result of their oxidation or reduction state.

Reduction: the gain of electrons by an atom or molecule.

Respiration: the oxidation of food/fuel to generate energy in the form of ATP.

Respiratory chain: also known as the electron transport chain (ETC), the series of complexes embedded in the bacterial membrane and inner mitochondrial membrane that pass electrons derived from fuel from one to the next. The energy released by the transport of electrons is used to pump protons across the membrane.

RNA: ribonucleic acid; includes messenger RNA (an exact copy of the DNA sequence in an individual gene, and moves to the cytoplasm); ribosomal RNA (forms part of the ribosomes, the protein-building factories); and transfer RNA (an adapter that couples a genetic code to a particular amino acid).

Sexual reproduction: reproduction brought by the fusion of two sex cells, each containing a random assortment of half the parental genes, to give the resulting embryo an equal number of genes from both parents.

Symbiosis: a mutually beneficial relationship between two.

Transcription factor: a protein that binds to a DNA sequence, signaling the transcription of that gene into an RNA copy (the first step in protein synthesis).

Tricarboxylic acid (TCA) cycle: also known as the Krebs cycle or the citric acid cycle; a metabolic pathway in the mitochondria that converts carbohydrates, fats, and proteins into energy compounds (NADH and $FADH_2$), which then enter the electron transport chain to ultimately create ATP.

Uncoupling: disconnecting oxidative phosphorylation from ATP production; instead, the proton gradient is dissipated by protons passing back through membrane pores (instead of ATPase), creating heat.

Uncoupling agent: any compound that disconnects oxidative phosphorylation from ATP production by dissipating the proton gradient.

Uncoupling protein: a channel in the membrane that allows protons to flow back through the membrane, dissipating the proton gradient as heat.

Uniparental inheritance: the inheritance of mitochondria from only one of two parents, specifically the mother.

BIBLIOGRAPHY

Chapter One

Althoff T, et al. Arrangement of electron transport chain components in bovine mitochondrial supercomplex I1III2IV1. EMBO J. 2011 Sep 9;30(22):4652–64. doi:10.1038/emboj.2011.324.

Ames BN, Shigenaga MK, Hagen TM. Oxidants, antioxidants, and the degenerative diseases of aging. Proc Natl Acad Sci USA. 1993 Sep 1; 90(17):7915–22.

Ames BN, Shigenaga MK, Hagen TM. Mitochondrial decay in aging. Biochim Biophys Acta. 1995 May 24;1271(1):165–70. doi:10.1016/0925-4439(95)00024-X

Aw TY, Jones DP. Nutrient supply and mitochondrial function. Annu Rev Nutr. 1989 Jul; 9:229–51. doi:10.1146/annurev.nu.09.070189.001305.

Bagh MB, et al. Age-related oxidative decline of mitochondrial functions in rat brain is prevented by long term oral antioxidant supplementation. Biogerontology. 2010 Sep 21; 12(2):119–31. doi:10.1007/s10522-010-9301-8.

Blackstone NW. Why did eukaryotes evolve only once? Genetic and energetic aspects of conflict and conflict mediation. Philos Trans R Soc Lond B Biol Sci. 2013 Jul 19;368(1622):20120266. doi:10.1098/rstb.2012.0266.

Brookes PS, et al. Calcium, ATP, and ROS: a mitochondrial love-hate triangle. Am J Physiol Cell Physiol. 2004 Oct;287(4):C817–33. doi:10.1152/ajpcell.00139.2004.

Bua EA, et al. Mitochondrial abnormalities are more frequent in muscles undergoing sarcopenia. J Appl Physiol (1985). 2002 Jun;92(6):2617–24. doi:10.1152/japplphysiol.01102.2001.

Buist R. Elevated xenobiotics, lactate and pyruvate in C.F.S. patients. J Orthomol Med. 1989; 4:170–2.

Cavalli LR, et al. Mutagenesis, tumorigenicity, and apoptosis: are the mitochondria involved? Mutat Res. 1998;398:19–26.

Chautan M, et al. Interdigital cell death can occur through a necrotic and caspase-independent pathway. Curr Biol. 1999 Sep 9;9(17):967–70. doi:10.1016/S0960-9822(99)80425-4.

Chiang SC, et al. Mitochondrial protein-linked DNA breaks perturb mitochondrial gene transcription and trigger free radical–induced DNA damage. Sci Adv. 2017 Apr 28;3(4): e1602506. doi:10.1126/sciadv.1602506.

Chinnery PF, Hudson G. Mitochondrial genetics. Br Med Bull. 2013;106:135–59. Epub 2013 May 22. doi:10.1093/bmb/ldt017.

Cohen BH, Gold DR. Mitochondrial cytopathy in adults: what we know so far. Cleve Clin J Med. 2001 Jul;68(7):625–26, 629–42.

Conley KE, et al. Ageing, muscle properties and maximal O2 uptake rate in humans. J Physiol. 2000 Jul 1;526(Pt 1):211–17. doi:10.1111/j.1469-7793.2000.00211.x.

Cooper GM. The cell: a molecular approach. 2nd ed. Sunderland, MA: Sinauer Associates; 2000.

Copeland WC, Longley MJ. Mitochondrial genome maintenance in health and disease. DNA Repair (Amst). 2014 Jul;19:190–8. Epub Apr 26. doi:10.1016/j.dnarep.2014.03.010.

Corral-Debrinski M, et al. Association of mitochondrial DNA damage with aging and coronary atherosclerotic heart disease. Mutat Res. 1992 Sep;275(3–6):169–80.

Croteau DL, Bohr VA. Repair of oxidative damage to nuclear and mitochondrial DNA in mammalian cells. J Biol Chem. 1997 Oct 10;272:25409–12. doi:10.1074/jbc.272.41.25409.

Einat H, Yuan P, Manji HK. Increased anxiety-like behaviors and mitochondrial dysfunction in mice with targeted mutation of the Bcl-2 gene: further support for the involvement of mitochondrial function in anxiety disorders. Behav Brain Res. 2005 Aug 10;165:172–80. doi:10.1016/j.bbr.2005.06.012.

Fattal O, et al. Review of the literature on major mental disorders in adult patients with mitochondrial diseases. Psychosomatics. 2006 Jan-Feb;47(1):1–7. doi:10.1176/appi.psy .47.1.1.

Fontaine E, et al. Regulation of the permeability transition pore in skeletal muscle mitochondria. J Biol Chem. 1998 May 15;273:12662–8. doi:10.1074/jbc.273.20.12662.

Fosslien E. Mitochondrial medicine — molecular pathology of defective oxidative phosphorylation. Ann Clin Lab Sci. 2001 Jan;31(1):25–67.

Fulle S, et al. Specific oxidative alterations in vastus lateralis muscle of patients with the diagnosis of chronic fatigue syndrome. Free Radic Biol Med. 2000;29:1252–9.

Garrett RH, Grisham CM. Biochemistry. Boston: Brooks/Cole; 2010.

Giles RE, et al. Maternal inheritance of human mitochondrial DNA. Proc Natl Acad Sci USA. 1980 Nov;77(11):6715–9.

Gill T, Levine AD. Mitochondrial derived hydrogen peroxide selectively enhances T cell receptor-initiated signal transduction. J Biol Chem. 2013 Sep 6;288(36):26246–55. Epub 2013 Jul 23. doi:10.1074/jbc.M113.476895.

Gray MW, Burger G, Lang BF. Mitochondrial evolution. Science. 1999 Mar 5;283(5407):1476–81.

Hagen TM, Wehr CM, Ames BN. Mitochondrial decay in aging. Reversal through supplementation of acetyl-L-carnitine and N-tert-butyl-alpha-phenyl-nitrone. Ann N Y Acad Sci. 1998 Nov 20;854:214–23.

Hengartner MO. The biochemistry of apoptosis. Nature. 2000;407(6805):770–6. doi:10.1038 /35037710.

Hirst J. Mitochondrial complex I. Annu Rev Biochem. 2013;82:551–75. Epub 2013 Mar 18. doi:10.1146/annurev-biochem-070511-103700.

Ip SW, et al. Capsaicin induces apoptosis in SCC-4 human tongue cancer cells through mitochondria-dependent and -independent pathways. Environ Toxicol. 2012 May; 27(6):332–41. Oct 5. doi:10.1002/tox.20646.

Javadov S, Kuznetsov A. Mitochondrial permeability transition and cell death: the role of cyclophilin d. Front Physiol. 2013 Apr 11;4:76. doi:10.3389/fphys.2013.00076.

Joza N, et al. Essential role of the mitochondrial apoptosis-inducing factor in programmed cell death. Nature 2001 Mar 29;410(6828):549–54. doi:10.1038/35069004.

Karbowski M, Youle RJ. Dynamics of mitochondrial morphology in healthy cells and during apoptosis. Cell Death Differ. 2003 Aug;10(8):870–80. doi:10.1038/sj.cdd.4401260.

Karp, Gerald. Cell and molecular biology. 5th ed. Hoboken, NJ: John Wiley & Sons; 2008.

Koike K. Molecular basis of hepatitis C virus-associated hepatocarcinogenesis: lessons from animal model studies. Clin Gastroenterol Hepatol. 2005 Oct;3(10 Suppl 2):S132–S135. doi:10.1016/S1542-3565(05)00700-7.

Kopsidas G, et al. An age-associated correlation between cellular bioenergy decline and mtDNA rearrangements in human skeletal muscle. Mutat Res. 1998 Oct 12; 421(1):27–36. doi:10.1016/S0027-5107(98)00150-X.

Ku HH, Brunk UT, Sohal RS. Relationship between mitochondrial superoxide and hydrogen peroxide production and longevity of mammalian species. Free Radic Biol Med. 1993 Dec;15(6):621–7.

Lagouge M, Larsson NG. The role of mitochondrial DNA mutations and free radicals in disease and ageing. J Intern Med. 2013 Jun;273(6):529–43. Epub 2013 Mar 7.

Lane, N. Power, sex, suicide: mitochondria and the meaning of life. New York: Oxford University Press; 2005.

Lane N. Bioenergetic constraints on the evolution of complex life. Cold Spring Harb Perspect Biol. 2014 May 1;6(5):a015982. doi:10.1101/cshperspect.a015982.

Lang BF, et al. An ancestral mitochondrial DNA resembling a eubacterial genome in miniature. Nature. 1997 May 29;387(6632):493–7. doi:10.1038/387493a0.

Lanza IR, Sreekumaran Nair K. Regulation of skeletal muscle mitochondrial function: genes to proteins. Acta Physiol (Oxf). 2010 Aug;199(4):529–47. doi:10.1111/j.1748-1716.2010 .02124.x.

Lapuente-Brun E, et al. Supercomplex assembly determines electron flux in the mitochondrial electron transport chain. Science. 2013 Jun 28;340(6140):1567–70. doi:10.1126/science .1230381.

Lieber CS, et al. Model of nonalcoholic steatohepatitis. Am J Clin Nutr. 2004 Mar;79(3):502–9.

Linnane AW, et al. Mitochondrial DNA mutations as an important contributor to aging and degenerative diseases. Lancet. 1989 Mar 25;1(8639):642–5. doi:10.1016/S0140-6736 (89)92145-4.

Linnane AW, et al. The universality of bioenergetic disease and amelioration with redox therapy. Biochim Biophys Acta. 1995 May 24;1271(1):191–4. doi:10.1016/0925-4439(95)00027-2.

Linnane AW, Kovalenko S, Gingold EB. The universality of bioenergetic disease. Age-associated cellular bioenergetic degradation and amelioration therapy. Ann N Y Acad Sci. 1998 Nov 20;854:202–13. doi:10.1111/j.1749-6632.1998.tb09903.x.

Liu J, et al. Delaying brain mitochondrial decay and aging with mitochondrial antioxidants and metabolites. Ann N Y Acad Sci. 2002 Apr;959:133–66. doi:10.1111/j.1749-6632.2002 .tb02090.x.

Luft R, et al. A case of severe hypermetabolism of nonthyroid origin with a defect in the maintenance of mitochondrial respiratory control: a correlated clinical, biochemical, and morphological study. J Clin Invest. 1962;41:1776–804.

Manczak M, et al. Mitochondria-targeted antioxidants protect against amyloid-beta toxicity in Alzheimer's disease neurons. J Alzheimers Dis. 2010;20 Suppl 2:S609–S631. doi:10.3233 /JAD-2010-100564.

Merry TL, Ristow M. Do antioxidant supplements interfere with skeletal muscle adaptation to exercise training? J Physiol. 2016 Sep 15;594(18):5135–47. doi:10.1113/JP270654.

Michikawa Y, et al. Aging-dependent large accumulation of point mutations in the human mtDNA control region for replication. Science. 1999 Oct 22;286(5440):774–9. doi:10.1126 /science.286.5440.774.

Mirisola MG, Longo VD. A radical signal activates the epigenetic regulation of longevity. Cell Metab. 2013 Jun 4;17(6):812–3. doi:10.1016/j.cmet.2013.05.015.

Murphy MP, Smith RA. Targeting antioxidants to mitochondria by conjugation to lipophilic cations. Annu Rev Pharmacol Toxicol. 2007;47:629–56. doi:10.1146/annurev.pharmtox .47.120505.105110.

Murray RK, et al. Harper's Illustrated Biochemistry. Hoboken, NJ: Lange Medical Books/ McGraw Hill; 2003.

Newmeyer DD, Ferguson-Miller S. Mitochondria: releasing power for life and unleashing the machineries of death. Cell. 2003 Feb 21;112(4):481–90. doi:10.1016/S0092-8674(03)00116-8.

Oelkrug R, et al. Brown fat in a protoendothermic mammal fuels eutherian evolution. Nat Commun. 2013 Jul 16;4:2140. doi:10.1038/ncomms3140.

Olsen LF, Issinger OG, Guerra B. The Yin and Yang of redox regulation. Redox Rep. 2013;18(6):245–52. doi:10.1179/1351000213Y.0000000059.

Ozawa T. Genetic and functional changes in mitochondria associated with aging. Physiol Rev. 1997 Apr 1;77(2):425–64.

Park JH, Niermann KJ, Olsen N. Evidence for metabolic abnormalities in the muscles of patients with fibromyalgia. Curr Rheumatol Rep. 2000;2(2):131–40.

Puddu P, et al. Mitochondrial dysfunction as an initiating event in atherogenesis: a plausible hypothesis. Cardiology. 2005;103(3):137–141. doi:10.1159/000083440.

Ricci JE, et al. Disruption of mitochondrial function during apoptosis is mediated by caspase cleavage of the p75 subunit of complex I of the electron transport chain. Cell. 2004 Jun 11;117(6):773–86. doi:10.1016/j.cell.2004.05.008.

Richter C, et al. Control of apoptosis by the cellular ATP level. 1996 Jan 8;FEBS Lett 378(2):107–10. doi:10.1016/0014-5793(95)01431-4.

Samsel A, Seneff S. Glyphosate, pathways to modern diseases II: celiac sprue and gluten intolerance. Interdiscip Toxicol. 2013 Dec;6(4):159–84. doi:10.2478/intox-2013-0026.

Sato M, Sato K. Maternal inheritance of mitochondrial DNA by diverse mechanisms to eliminate paternal mitochondrial DNA. Biochim Biophys Acta. 2013 Aug;1833(8):1979–84. Epub 2013 Mar 21.

Savitha S, et al. Efficacy of levo carnitine and alpha lipoic acid in ameliorating the decline in mitochondrial enzymes during aging. Clin Nutr. 2005 Oct;24(5):794–800. doi:10.1016/j .clnu.2005.04.005.

Schroeder EA, Raimundo N, Shadel GS. Epigenetic silencing mediates mitochondria stress-induced longevity. Cell Metab. 2013 Jun 4;17(6):954–64. doi:10.1016/j.cmet.2013.04.003.

Skulachev VP, Longo VD. Aging as a mitochondria-mediated atavistic program: can aging be switched off? Ann NY Acad Sci. 2005 Dec;1057:145–64. doi:10.1196/annals.1356.009.

Smith RA, et al. Mitochondria-targeted antioxidants in the treatment of disease. Ann N Y Acad Sci. 2008 Dec;1147:105–11. doi:10.1196/annals.1427.003.

Sohal RS, Sohal BH, Orr WC. Mitochondrial superoxide and hydrogen peroxide generation, protein oxidative damage, and longevity in different species of flies. Free Radic Biol Med. 1995 Oct;19(4):499–504. doi:10.1016/0891-5849(95)00037-X.

Stavrovskaya IG, Kristal BS. The powerhouse takes control of the cell: is the mitochondrial permeability transition a viable therapeutic target against neuronal dysfunction and death? Free Radic Biol Med. 2005 Mar 15;38(6):687–97. doi:10.1016/j.freeradbiomed.2004.11.032.

Stork C, Renshaw PF. Mitochondrial dysfunction in bipolar disorder: evidence from magnetic resonance spectroscopy research. Mol Psychiatry. 2005 Oct;10(10):900–19. doi:10.1038/sj.mp.4001711.

Susin SA, et al. Mitochondria as regulators of apoptosis: doubt no more. Biochim Biophys Acta. 1998 Aug 10;1366(1–2):151–65.doi:10.1016/S0005-2728(98)00110-8.

Tait SW, Green DR. Mitochondrial regulation of cell death. Cold Spring Harb Perspect Biol. 2013 Sep 1;5(9):pii:a008706. doi:10.1101/cshperspect.a008706.

Turker MS. Somatic cell mutations: can they provide a link between aging and cancer? Mech Aging Dev. 2000 Aug 15;117(1–3):1–19. doi:10.1016/S0047-6374(00)00133-0.

Van Raamsdonk JM. Levels and location are crucial in determining the effect of ROS on lifespan. Worm. 2015 Oct–Dec;4(4):e1094607. doi:10.1080/21624054.2015.1094607.

Vartak R, Porras CA, Bai Y. Respiratory supercomplexes: structure, function and assembly. Protein Cell. 2013 Aug;4(8):582–90. Epub 2013 Jul 5. doi:10.1007/s13238-013-3032-y.

Wallace DC. A mitochondrial paradigm of metabolic and degenerative diseases, aging, and cancer: a dawn for evolutionary medicine. Annu Rev Genet. 2005;39:359–407. doi:10.1146 /annurev.genet.39.110304.095751.

Wallace DC. Why do we still have a maternally inherited mitochondrial DNA? Insights from evolutionary medicine. Annu Rev Biochem. 2007;76:781–821. doi:10.1146/annurev .biochem.76.081205.150955.

Wallace DC. A mitochondrial bioenergetic etiology of disease. J Clin Invest. 2013 Apr;123(4): 1405–12. Epub 2013 Apr 1. doi:10.1172/JCI61398.

Wallace DC, et al. Mitochondrial DNA mutations in human degenerative diseases and aging. Biochim Biophys Acta. 1995 May 24;1271(1):141–51. doi:10.1016/0925-4439(95)00021-U.

Wang CH, et al. Oxidative stress response elicited by mitochondrial dysfunction: implication in the pathophysiology of aging. Exp Biol Med (Maywood). 2013 May;238(5):450–60. doi:10.1177/1535370213493069.

Wei YH, Kao SH, Lee HC. Simultaneous increase of mitochondrial DNA deletions and lipid peroxidation in human aging. Proc NY Acad Sci. 1996 Jun 15;786:24–43. doi:10.1111 /j.1749-6632.1996.tb39049.x.

West IC. Radicals and oxidative stress in diabetes. Diabet Med. 2000 Mar;17(3):171–80. doi:10.1046/j.1464-5491.2000.00259.x.

Wolvetang EJ, et al. Mitochondrial respiratory chain inhibitors induce apoptosis. 1994 Feb 14; 339(1–2):40–4. doi:10.1016/0014-5793(94)80380-3.

Wookieepedia. Midi-chlorian [Internet]. [Cited 2011 Dec 27]. http://starwars.wikia.com/wiki /Midi-chlorian.

Yunus MB, Kalyan-Raman UP, Kalyan-Raman K. Primary fibromyalgia syndrome and myofascial pain syndrome: clinical features and muscle pathology. Arch Phys Med Rehabil. 1988 Jun;69(6):451–4.

Zhang M, Mileykovskaya E, Dowhan W. Gluing the respiratory chain together: cardiolipin is required for supercomplex formation in the inner mitochondrial/membrane. J Biol Chem. 2002 Nov 15;277(46):43553–6. doi:10.1074/jbc.C200551200.

Chapter Two

Hirst J. Mitochondrial complex I. Annu Rev Biochem. 2013;82:551–75. Epub 2013 Mar 18. doi:10.1146/annurev-biochem-070511-103700.

Hwang AB, Jeong DE, Lee SJ. Mitochondria and organismal longevity. Curr Genomics. 2012 Nov;13(7):519–32. doi:10.2174/138920212803251427.

Lane, N. Power, sex, suicide: mitochondria and the meaning of life. New York: Oxford University Press; 2005.

Munro D, et al. Low hydrogen peroxide production in mitochondria of the long-lived Arctica islandica: underlying mechanisms for slow aging. Aging Cell. 2013 Aug;12(4):584–92. Epub 2013 May 6. doi:10.1111/acel.12082.

Sinatra ST. The Sinatra solution: metabolic cardiology. Laguna Beach, CA: Basic Health Publications, Inc; 2011.

Wallace DC. Mitochondrial genetics: a paradigm for aging and degenerative diseases? Science. 1992 May 1;256(5057):628–32. doi:10.1126/science.1533953.

Wallace DC. A mitochondrial bioenergetic etiology of disease. J Clin Invest. 2013 Apr;123(4): 1405–12. Epub 2013 Apr 1. doi:10.1172/JCI61398.

The Role of Mitochondria in Cardiovascular Disease

Aon MA. Mitochondrial dysfunction, alternans, and arrhythmias. Front Physiol. 2013 Apr 19;4:83.

Buja LM. The pathobiology of acute coronary syndromes: clinical implications and central role of the mitochondria. Tex Heart Inst J. 2013;40(3):221–8.

Gorenkova N, et al. Conformational change of mitochondrial complex I increases ROS sensitivity during ischaemia. Antioxid Redox Signal. 2013 Oct;19(13):1459–68. Epub 2013 Feb 18. doi:10.1089/ars.2012.4698.

Li H, Horke S, Förstermann U. Oxidative stress in vascular disease and its pharmacological prevention. Trends Pharmacol Sci. 2013 Jun;34(6):313–9. Epub 2013 Apr 19. doi:10.1016/j.tips.2013.03.007.

Lonnrot K, et al. Control of arterial tone after long-term coenzyme Q10 supplementation in senescent rats. Brit J Pharmacol. 1998 Aug;124(7):1500–6. doi:10.1038/sj.bjp.0701970.

Karamanlidis G, et al. Defective DNA replication impairs mitochondrial biogenesis in human failing hearts. Circ Res. 2010 May 14;106(9):1541–8. doi:10.1161/CIRCRESAHA.109.212753.

Knight-Lozano CA, et al. Cigarette smoke exposure and hypercholesterolemia increase mitochondrial damage in cardiovascular tissues. Circulation. 2002 Feb 19;105(7):849–54. doi:10.1161/hc0702.103977.

Madamanchi NR, Runge MS. Mitochondrial dysfunction in atherosclerosis. Circ Res. 2007 Mar 2;100(4):460–73.

Mercer JR. Mitochondrial bioenergetics and therapeutic intervention in cardiovascular disease. Pharmacol Ther. 2014 Jan;141(1):13–20. Epub. doi:10.1016/j.pharmthera.2013.07.011.

Montaigne D, et al. Mitochondrial dysfunction as an arrhythmogenic substrate: a translational proof-of-concept study in patients with metabolic syndrome in whom post-operative atrial fibrillation develops. J Am Coll Cardiol. 2013 Oct 15;62(16):1466–73. Epub 2013 May 1. doi:10.1016/j.jacc.2013.03.061.

Morales CR, et al. Oxidative stress and autophagy in cardiovascular homeostasis. Antioxid Redox Signal. 2014 Jan 20;20(3):507–518. Epub 2013 May 5. doi:10.1089/ars.2013.5359.

Nazarewicz RR, Dikalov SI. Mitochondrial ROS in the pro-hypertensive immune response. Am J Physiol Regul Integr Comp Physiol. 2013 May 8;305:R98–100. Epub. doi:10.1152/ajpregu.00208.2013.

Oeseburg H, et al. Bradykinin protects against oxidative stress-induced endothelial cell senescence. Hypertension. 2009 Feb;53(Part 2):417–22. doi:10.1161/HYPERTENSIONAHA.108.123729.

Schleicher M, et al. Prohibitin-1 maintains the angiogenic capacity of endothelial cells by regulating mitochondrial function and senescence. J Cell Biol. 2008 Jan 14;180(1):101–12. doi:10.1083/jcb.200706072.

Schriewer JM, et al. ROS-mediated PARP activity undermines mitochondrial function after permeability transition pore opening during myocardial ischemia-reperfusion. J Am Heart Assoc. 2013 Apr 18;2(2):e000159. doi:10.1161/JAHA.113.000159.

Stride N, et al. Impaired mitochondrial function in chronically ischemic human heart. Am J Physiol Heart Circ Physiol. 2013 Mar 29. Epub. doi:10.1152/ajpheart.00991.2012.

Wallace DC. A mitochondrial paradigm of metabolic and degenerative diseases, aging, and cancer: a dawn for evolutionary medicine. Annu Rev Genet. 2005;39:359–407. doi:10.1146/annurev.genet.39.110304.095751.

Yang Z, et al. Prenatal environmental tobacco smoke exposure promotes adult atherogenesis and mitochondrial damage in apolipoprotein E-/- mice fed a chow diet. Circulation. 2004 Dec 14; 110(24):3715–20. doi:10.1161/01.CIR.0000149747.82157.01.

Yang Z, et al. The role of tobacco smoke induced mitochondrial damage in vascular dysfunction and atherosclerosis. Mutat Res. 2007 Aug 1;621(1–2):61–74. doi:10.1016/j.mrfmmm.2007.02.010.

Understanding Smooth Muscles

Chitaley K, Weber DS, Webb RC. RhoA/Rho-kinase, vascular changes and hypertension. Curr Hypertension Rep. 2001;3:139–144. doi:10.1007/s11906-001-0028-4

Feletou M, Vanhoutte PM. Endothelium-dependent hyperpolarization of vascular smooth muscle cells. Acta Pharmacol Sin. 2000 Jan;21(1):1–18.

Fukata Y, Mutsuki A, Kaibuchi K. Rho-Rho-kinase pathway in smooth muscle contraction and cytoskeletal reorganization of non-muscle cells. Trends Physiol Sci. 2001 Jan;22(1):32–9. doi:10.1016/S0165-6147(00)01596-0.

Jin L, et al. Inhibition of the tonic contraction in the treatment of erectile dysfunction. Exp Opin Ther Targets. 2003;7(2):265–76. doi:10.1517/14728222.7.2.265.

Kao CY, Carsten ME, editors. Cellular Aspects of Smooth Muscle Function. New York: Cambridge Univ. Press, 1997. Chapter 5, Mechanics of smooth muscle contraction; p. 169–208.

Kohlhaas M, Maack C. Calcium release microdomains and mitochondria. Cardiovasc Res. 2013 Feb 14;98:259–68. Epub. doi:10.1093/cvr/cvt032.

Lanza IR, Sreekumaran Nair K. Regulation of skeletal muscle mitochondrial function: genes to proteins. Acta Physiol (Oxf). 2010 Aug;199(4):529–47. doi:10.1111/j.1748-1716.2010 .02124.x.

Li M, et al. High glucose concentrations induce oxidative damage to mitochondrial DNA in explanted vascular smooth muscle cells. Exp Biol Med. 2001 Jan 1;226(5):450–7. doi:10.1177 /153537020122600510.

Mehta S, Webb RC, Dorrance AM. The pathophysiology of ischemic stroke: a neuronal and vascular perspective. J Med Sci. 2002;22:53–62.

Mills TM, et al. Inhibition of tonic contraction—a novel way to approach erectile dysfunction? J Androl. 2002 Sep 10;23(5):S5–S9. doi:10.1002/j.1939-4640.2002.tb02294.x.

Mitchell BM, Chitaley KC, Webb RC. Vascular smooth muscle contraction and relaxation. In: Izzo JL, Black HR, editors. Hypertension primer: the essentials of high blood pressure. Dallas, TX: Am. Heart Assoc.; 2003, p. 97–99.

Morgan KG. The role of calcium in the control of vascular tone as assessed by the Ca2+ indicator aequorin. Cardiovasc Drugs Ther. 1990 Oct;4(5):1355–62.

Ridley A. Rho: theme and variations. Curr Biol 1996;6(10):1256–64. doi:10.1016/S0960-9822 (02)70711-2.

Sah VP, et al. The role of Rho in G protein-coupled receptor signal transduction. Annu Rev Pharmacol Toxicol. 2000;40:459–89. doi:10.1146/annurev.pharmtox.40.1.459.

Solaro RJ. Myosin light chain phosphatase: a Cinderella of cellular signaling. Circ Res. 2000 Aug 4;87:173–5. doi:10.1161/01.RES.87.3.173.

Somlyo AP, Somlyo AV. From pharmacomechanical coupling to G-proteins and myosin phosphatase. Acta Physiol Scand. 1998 Dec;164(4):437–48. doi:10.1046/j.1365-201X .1998.00454.x.

Somlyo AP, Somlyo AV. Signal transduction by G-proteins, Rho-kinase and protein phosphatase to smooth muscle and non-muscle myosin II. J Physiol. 2000;522(Pt 2):177–85. doi:10.1111/j.1469-7793.2000.t01-2-00177.x.

Somlyo AP, et al. Pharmacomechanical coupling: the role of calcium, G-proteins, kinases and phosphatases. Rev Physiol Biochem Pharmacol. 1999;134:201–34.

Uehata M, et al. Calcium sensitization of smooth muscle mediated by a Rho-associated protein kinase in hypertension. Nature. 1997;389:990–4. doi:10.1038/40187.

Woodrum DA, Brophy CM. The paradox of smooth muscle physiology. Mol Cell Endocrinol. 2001;177(1–2):135–43. doi:10.1016/S0303-7207(01)00407-5.

The Role of Mitochondria in the Nervous System, Brain, and Cognitive Health

Allen KL, et al. Changes of respiratory chain activity in mitochondrial and synaptosomal fractions isolated from the gerbil brain after graded ischaemia. J Neurochem. 1995 May;64(5):2222–9. doi:10.1046/j.1471-4159.1995.64052222.x.

Ankarcrona M, et al. Glutamate-induced neuronal death: a succession of necrosis or apoptosis depending on mitochondrial function. Neuron. 1995 Oct;15(4):961–73. doi:10.1016 /0896-6273(95)90186-8.

Barbiroli B, et al. Coenzyme Q10 improves mitochondrial respiration in patients with mitochondrial cytopathies. An in vivo study on brain and skeletal muscle by phosphorous magnetic resonance spectroscopy. Cell Molec Biol. 1997;43:741–9.

Beal MF. Aging, energy, and oxidative stress in neurodegenerative diseases. Ann Neurol. 1995 Sep;38(3):357–66. doi:10.1002/ana.410380304.

Beal MF, et al. Coenzyme Q10 and nicotinamide block striatal lesions produced by the mitochondrial toxin malonate. Ann Neurol. 1994;36(6):882–8. doi:10.1002/ana.410360613.

Beal MF, et al. Coenzyme Q10 attenuates the 1-methyl-4-phenyl-1,2,3,6-tetrahydropyridine (MPTP) induced loss of striatal dopamine and dopaminergic axons in aged mice. Brain Res. 1998 Feb;783(1):109–14. doi:10.1016/S0006-8993(97)01192-X.

Bendahan D, et al. 31P NMR spectroscopy and ergometer exercise test as evidence for muscle oxidative performance improvement with coenzyme Q in mitochondrial myopathies. Neurology. 1992;42(6):1203–8.

Berchtold NC, et al. Brain gene expression patterns differentiate mild cognitive impairment from normal aged and Alzheimer's disease. Neurobiol Aging. 2014 Sep;35(9):1961–72. Epub 2014 Apr 2. doi:10.1016/j.neurobiolaging.2014.03.031.

Bolanos JP, et al. Nitric oxide-mediated mitochondrial damage in the brain: mechanisms and implications for neurodegenerative diseases. J Neurochem.1997 Jun;68(6):2227–40. doi:10.1046/j.1471-4159.1997.68062227.x.

Bozner P, et al. The amyloid β protein induces oxidative damage of mitochondrial DNA. J Neuropathol Exp Neurol. 1997;56:1356–62. doi:10.1097/00005072-199712000-00010.

Brookes PS, et al. Peroxynitrite and brain mitochondria: evidence for increased proton leak. J Neurochem. 1998;70(No 5):2195–02.

Casley CS, et al. Beta-amyloid inhibits integrated mitochondrial respiration and key enzyme activities. J Neurochem. 2002 Jan;80(1):91–100. doi:10.1046/j.0022-3042.2001.00681.x.

Cassarino DS, et al. An evaluation of the role of mitochondria in neurodegenerative diseases: mitochondrial mutations and oxidative pathology, protective nuclear responses, and cell death in neurodegeneration. Brain Res Brain Res Rev. 1999 Jan;29(1):1–25. doi:10.1016 /S0165-0173(98)00046-0.

Chaturvedi RK, Flint Beal M. Mitochondrial diseases of the brain. Free Radic Biol Med. 2013 Oct;63:1–29. Epub Apr 5. doi:10.1016/j.freeradbiomed.2013.03.018.

de Moura MB, dos Santos LS, Van Houten B. Mitochondrial dysfunction in neurodegenerative diseases and cancer. Environ Mol Mutagen. 2010 Jun;51(5):391–405. doi:10.1002/em.20575.

Favit A, et al. Ubiquinone protects cultured neurons against spontaneous and excitotoxin-induced degeneration. J Cereb Blood Flow Metab.1992;12(No 4):638–45.

Fiskum G, Murphy AN, Beal MF. Mitochondria in neurodegeneration: acute ischemia and chronic neurodegenerative diseases. J Cereb Blood Flow Metab. 1999 Apr;19(4):351–69. doi:10.1097/00004647-199904000-00001.

Kuroda S, Siesjo BK. Reperfusion damage following focal ischemia: pathophysiology and therapeutic windows. Clin Neurosci. 1997;4(4):199–212.

Leist M, Nicotera P. Apoptosis, excitotoxicity, and neuropathology. Exp Cell Res. 1998;239(2): 183–201. doi:10.1006/excr.1997.4026.

Liu J, et al. Memory loss in old rats is associated with brain mitochondrial decay and RNA/ DNA oxidation: partial reversal by feeding acetyl-L-carnitine and/or R-alpha-lipoic acid. Proc Natl Acad Sci U S A. 2002 Feb 19;99(4):2356–61. doi:10.1073/pnas.261709299.

Love S. Oxidative stress in brain ischemia. Brain Pathol. 1999 Jan;9(1):119–31. doi:10.1111 /j.1750-3639.1999.tb00214.x.

Matsumoto S, et al. Blockade of the mitochondrial permeability transition pore diminishes infarct size in the rat after transient middle cerebral artery occlusion. J Cereb Blood Flow Metab. 1999;19(No 7):736–41.

Matthews RT, et al. Coenzyme Q10 administration increases brain mitochondrial concentrations and exerts neuroprotective effects. Proc Natl Acad Sci U S A. 1998 Jul 21;95 (15):8892–7.

Mazzio E, et al. Effect of antioxidants on L-glutamate and N-methyl-4-phenylpyridinium ion induced-neurotoxicity in PC12 cells. Neurotoxicology. 2001;22:283–8.

Mecocci P, et al. Oxidative damage to mitochondrial DNA shows marked age-dependent increases in human brain. Ann Neurol. 1993 Oct;34(4):609–16. doi:10.1002/ana.410340416.

Mordente A, et al. Free radical production by activated haem proteins: protective effect of coenzyme Q. Molec Aspects Med. 1994;15(Suppl S109–S115).

Murphy AN, Fiskum G, Beal F. Mitochondria in neurodegeneration: bioenergetic function in cell life and death. J Cereb Blood Flow Metab. 1999;19(No 3):231–45.

Musumeci O, et al. Familial cerebellar ataxia with muscle coenzyme Q10 deficiency. Neurology. 2001 Apr 10;56(7):849–55.

Nam MK, et al. Essential roles of mitochondrial depolarization in neuron loss through microglial activation and attraction toward neurons. Brain Res. 2013 Apr 10;1505:75–85. Epub Feb 12. doi:10.1016/j.brainres.2013.02.005.

Novelli A, et al. Glutamate becomes neurotoxic via the N-methyl-D-aspartate receptor when intracellular energy levels are reduced. Brain Res.1988 Jun 7;451(1–2):205–12. doi:10.1016/0006-8993(88)90765-2.

Ristow M, et al. Frataxin activates mitochondrial energy conversion and oxidative phosphorylation. Proc Natl Acad Sci U S A. 2000;97(No 22):12239–43. doi:10.1073 /pnas.220403797.

Schon EA, Manfredi G. Neuronal degeneration and mitochondrial dysfunction. J Clin Invest. 2003 Feb;111(3):303–12. doi:10.1172/JCI17741.

Schulte EC, et al. Mitochondrial membrane protein associated neurodegenration: A novel variant of neurodegeneration with brain iron accumulation. Mov Disord. 2013 Feb;28(2):224–7. Epub 2012 Nov 19. doi:10.1002/mds.25256.

Schulz JB, et al. Neuroprotective strategies for treatment of lesions produced by mitochondrial toxins: implications for neurodegenerative diseases. Neuroscience 71. 1996;71(4):1043–48. doi:10.1016/0306-4522(95)00527-7.

Sobreira C, et al. Mitochondrial encephalomyopathy with coenzyme Q10 deficiency. Neurology. 1997 May;48(5):1238–43.

Sun T, et al. Motile axonal mitochondria contribute to the variability of presynaptic strength. Cell Rep. 2013 Aug 15;4(3):413–9. Epub 2013 Jul 23. doi:10.1016/j.celrep.2013.06.040.

Tatton WG, Chalmers-Redman RM. Mitochondria in neurodegenerative apoptosis: an opportunity for therapy? Ann Neurol. 1998;44(3 Suppl 1):S134–S141. doi:10.1002/ana .410440720.

Tatton WG, Olanow CW. Apoptosis in neurodegenerative diseases: the role of mitochondria. Biochim Biophys Acta. 1999 Feb 9;1410(2):195–213. doi:10.1016/S0005-2728(98)00167-4.

Turner C, Schapira AH. Mitochondrial dysfunction in neurodegenerative disorders and ageing. Adv Exp Med Biol. 2001;487:229–51.

Veitch K et al. Global ischemia induces a biphasic response of the mitochondrial respiratory chain. Anoxic pre-perfusion protects against ischaemic damage. Biochem J. 1992 Feb 1;281(Pt 3):709–15.

Volpe M, Cosentino F. Abnormalities of endothelial function in the pathogenesis of stroke: the importance of endothelin. J Cardiovasc Pharmacol. 2000;35(4 Suppl 2):S45–S48.

Alzheimer's Disease: Don't Forget the Mitochondria!

Berger A. The Alzheimer's antidote: Using a low-carb, high-fat diet to fight Alzheimer's disease, memory loss, and cognitive decline. White River Junction, VT: Chelsea Green Publishing, 2017.

Blass JP. The mitochondrial spiral. An adequate cause of dementia in the Alzheimer's syndrome. Ann N Y Acad Sci. 2000;924:170–83. doi:10.1111/j.1749-6632.2000.tb05576.x.

Bonilla E, et al. Mitochondrial involvement in Alzheimer's disease. Biochim Biophys Acta. 1999 Feb 9;1410(2):171–82. doi:10.1016/S0005-2728(98)00165-0.

Brown AM, et al. Correlation of the clinical severity of Alzheimer's disease with an aberration in mitochondrial DNA (mtDNA). J Mol Neurosci. 2001 Feb;16(1):41–8. doi:10.1385/JMN:16:1:41.

Cavallucci V, Ferraina C, D'Amelio M. Key role of mitochondria in Alzheimer's disease synaptic dysfunction. Curr Pharm Des. 2013;19(36):6440–50. Epub 2013 Feb 13.

Chen JX, Yan SD. Amyloid-beta-induced mitochondrial dysfunction. J Alzheimers Dis. 2007 Sep;12(2):177–84. doi:10.3233/JAD-2007-12208.

Duboff B, Feany M, Götz J. Why size matters — balancing mitochondrial dynamics in Alzheimer's disease. Trends Neurosci. 2013 Jun;36(6):325–35. Epub 2013 Apr 11. doi:10.1016/j.tins.2013.03.002.

Gabuzda D, et al. Inhibition of energy metabolism alters the processing of amyloid precursor protein and induces a potentially amyloidogenic derivative. J Biol Chem. 1994 May 6;269(18):13623–8.

Harman D. A hypothesis on the pathogenesis of Alzheimer's disease. Ann N Y Acad Sci. 1996 Jun 15;786:152–68. doi:10.1111/j.1749-6632.1996.tb39059.x.

Hu H, et al. A mitocentric view of Alzheimer's disease. Mol Neurobiol. 2016 Oct 1. Epub ahead of print. doi:10.1007/s12035-016-0117-7.

Lustbader JW, et al. ABAD directly links Abeta to mitochondrial toxicity in Alzheimer's disease. Science. 2004 Apr 16;304(5669):448–52. doi:10.1126/science.1091230.

Mariani C, et al. Muscle biopsy in Alzheimer's disease: morphological and biochemical findings. Clin Neuropathol. 1991 Jul;10(4):171–6.

Mark RJ, et al. Amyloid b-peptide impairs glucose transport in hippocampal and cortical neurons: involvement of membrane lipid peroxidation. J Neurosci. 1997 Feb 1;17(3):1046–54.

Markesbery WR. Oxidative stress hypothesis in Alzheimer's disease. Free Radic Biol Med. 1997;23(1):134–47. doi:10.1016/S0891-5849(96)00629-6.

Markesbery WR. Oxidative alterations in Alzheimer's disease. Brain Pathol. 1999 Jan;9(1):133–46. doi:10.1111/j.1750-3639.1999.tb00215.x.

Muller WE, et al. Mitochondrial dysfunction: common final pathway in brain aging and Alzheimer's disease—therapeutic aspects. Mol Neurobiol. 2010 Jun;41(2–3):159–71. doi:10.1007/s12035-010-8141-5.

Munch G, et al. Alzheimer's disease — synergistic effects of glucose deficit, oxidative stress and advanced glycation endproducts. J Neural Transm (Vienna). 1998;105(4–5):439–61. doi:10.1007/s007020050069.

Nia SS, et al. New pathogenic variations of mitochondrial DNA in Alzheimer disease! [letter]. J Res Med Sci. 2013 Mar;18(3):269.

Nicotera P, Leist M, Manzo L. Neuronal cell death: a demise with different shapes. Trends Pharmacol Sci. 1999 Feb 1;20(2):46–51. doi:10.1016/S0165-6147(99)01304-8.

Ogawa M, et al. Altered energy metabolism in Alzheimer's disease. J Neurol Sci. 1996 Jul;139(1):78–82. doi:10.1016/0022-510X(96)00033-0.

Sery O, et al. Molecular mechanisms of neuropathological changes in Alzheimer's disease: a review. Folia Neuropathol. 2013;51(1):1–9. doi:10.5114/fn.2013.34190.

Smith MA, et al. Widespread peroxynitrite-mediated damage in Alzheimer's disease. J Neurosci 1997 Apr 15;17(8):2653–7.

Sochocka M, et al. Vascular oxidative stress and mitochondrial failure in the pathobiology of Alzheimer's disease: new approach to therapy. CNS Neurol Disord Drug Targets. 2013 Sep;12(6):870–81. Epub Feb 27. doi:10.2174/18715273113129990072.

Wang X, et al. Impaired balance of mitochondrial fission and fusion in Alzheimer's disease. J Neurosci. 2009 Jul 15;29(28):9090–103. doi:10.1523/JNEUROSCI.

Webster MT, et al. The effects of perturbed energy metabolism on the processing of amyloid precursor protein in PC12 cells. J Neural Transm. 1998 Nov;105(8–9):839–53. doi:10.1007/s007020050098.

Ying W. Deleterious network: a testable pathogenetic concept of Alzheimer's disease. Gerontology. 1997;43:242–53. doi:10.1159/000213856.

Overeating and Alzheimer's Disease

Adeghate E, Donath T, Adem A. Alzheimer disease and diabetes mellitus: do they have anything in common? Curr Alzheimer Res. 2013 Jul;10(6):609–17. Epub Apr 29. doi:10.2174/15672050113109990009.

Cetinkalp S, Simsir IY, Ertek S. Insulin resistance in brain and possible therapeutic approaches. Curr Vasc Pharmacol. 2014;12(4):553–64. Epub Apr 25. doi:10.2174/1570161112999140206130426.

Geda YE. Abstract 3431. Paper presented at: American Academy of Neurology (AAN) 64th Annual Meeting; 2012 Apr 21–28;. New Orleans, Louisiana.

Mastrogiacomo F, Bergeron C, Kish EJ. Brain alpha-ketoglutarate dehydrogenase complex activity in Alzheimer's disease. J Neurochem. 1993 Dec;61(6):2007–14. doi:10.1111/j.1471-4159.1993.tb07436.x.

Parkinson's Disease: Rethinking L-Dopa Therapy

Abou-Sleiman PM, Muqit MM, Wood NW. Expanding insights of mitochondrial dysfunction in Parkinson's disease. Nat Rev Neurosci. 2006 Mar;7(3):207–19. doi:10.1038/nrn1868.

Beal MF. Therapeutic approaches to mitochondrial dysfunction in Parkinson's disease. Parkinsonism Relat Disord. 2009 Dec;15 Suppl 3:S189–S194. doi:10.1016/S1353-8020(09)70812-0.

Beal MF, et al. Coenzyme Q10 attenuates the 1-methyl-4-phenyl-1,2,3,6-tetrahydropyridine (MPTP) induced loss of striatal dopamine and dopaminergic axons in aged mice. Brain Res. 1998 Feb;783(1):109–14. doi:10.1016/S0006-8993(97)01192-X.

Bender A, et al. TOM40 mediates mitochondrial dysfunction induced by α-synuclein accumulation in Parkinson's disease. PLoS One. 2013 Apr 23;8(4):e62277.

Berndt N, Holzhutter HG, Bulik S. Implications of enzyme deficiencies on the mitochondrial energy metabolism and ROS formation of neurons involved in rotenone-induced Parkinson's disease: A model-based analysis. FEBS J. 2013 Sep 12;280(20):5080–93. Epub 2013 Aug 13. doi:10.1111/febs.12480.

Dolle C, et al. Defective mitochondrial DNA homeostasis in the substantia nigra in Parkinson disease. Nat Commun. 2016 Nov 22;7:13548.

Ebadi M, et al. Ubiquinone (coenzyme q10) and mitochondria in oxidative stress of Parkinson's disease. 2001. Biol Signals Recept 10:224–53. doi:10.1038/ncomms13548.

Freeman D, et al. Alpha-synuclein induces lysosomal rupture and cathepsin dependent reactive oxygen species following endocytosis. PLoS One. 2013 Apr 25;8(4):e62143.

Haas RH, et al. Low platelet mitochondrial complex I and complex II/III activity in early untreated Parkinson's disease. Ann Neurol. 1995 Jun;37(6):714–22. doi:10.1002/ana.410370604.

Henchcliffe C, Beal MF. Mitochondrial biology and oxidative stress in Parkinson disease pathogenesis. Nat Clin Pract Neurol. 2008 Nov;4(11):600–9. doi:10.1038/ncpneuro0924.

Hosamani R, Muralidhara. Acute exposure of Drosophila melanogaster to paraquat causes oxidative stress and mitochondrial dysfunction. Arch Insect Biochem Physiol. 2013 May;83(1):25–40. Epub 2013 Apr 5.

Isobe C, Abe T, Terayama Y. Levels of reduced and oxidized coenzyme Q-10 and 8-hydroxy-2'-deoxyguanosine in the cerebrospinal fluid of patients living with Parkinson's disease demonstrate that mitochondrial oxidative damage and/or oxidative DNA damage contributes to the neurodegenerative process. Neurosci Lett. 2010 Jan 18;469(1):159–63. Epub 2009 Nov 26.

Lehmann S, Martins LM. Insights into mitochondrial quality control pathways and Parkinson's disease. J Mol Med (Berl). 2013 Jun;91(6):665–71. Epub May 4. doi:10.1007/s00109-013-1044-y.

Li DW, et al. α-lipoic acid protects dopaminergic neurons against MPP+-induced apoptosis by attenuating reactive oxygen species formation. Int J Mol Med. 2013 Jul;32(1):108–14. Epub Apr 24. doi:10.3892/ijmm.2013.1361.

Lin TK, et al. Mitochondrial dysfunction and biogenesis in the pathogenesis of Parkinson's disease. Chang Gung Med J. 2009 Nov–Dec;32(6):589–99.

Lodi R, et al. Antioxidant treatment improves in vivo cardiac and skeletal muscle bioenergetics in patients with Friedreich's ataxia. Ann Neurol. 2001 May 1;49(5):590–6. doi:10.1002/ana.1001.

Mena MA, et al. Neurotoxicity of levodopa on catecholamine-rich neurons. Mov Disord. 1992;7(1):23–31. doi:10.1002/mds.870070105.

Mizuno Y, et al. Role of mitochondria in the etiology and pathogenesis of Parkinson's disease. Biochima et Biophysica Acta. 1995 May 24;1271(1):265–74. doi:10.1016/0925-4439(95)00038-6.

Mizuno Y, et al. Mitochondrial dysfunction in Parkinson's disease. Ann Neurol. 1998 Sep;44 (3 Suppl 1):S99–S109.

Musumeci O, et al. Familial cerebellar ataxia with muscle coenzyme Q10 deficiency. Neurology. 2001 Apr 10;56(7):849–55.

Nakamura K. α-Synuclein and mitochondria: partners in crime? Neurotherapeutics. 2013 Jul;10(3):391–9. Epub Mar 20. doi:10.1007/s13311-013-0182-9.

Olanow CW, et al. The effect of deprenyl and levodopa on the progression of Parkinson's disease. Ann Neurol. Nov 1995;38(5):771–7. doi:10.1002/ana.410380512.

Perfeito R, Cunha-Oliveira T, Rego AC. Revisiting oxidative stress and mitochondrial dysfunction in the pathogenesis of Parkinson's disease — resemblance to the effect of amphetamine drugs of abuse. Free Radic Biol Med. 2012 Nov 1;53(9):1791–806. doi:10.1016/j.freeradbiomed.2012.08.569.

Przedborski S, Jackson-Lewis V, Fahn S. Antiparkinsonian therapies and brain mitochondrial complex I activity. Mov Disord. May 1995;10(3):312–7. doi:10.1002/mds.870100314.

Schapira AH, et al. Novel pharmacological targets for the treatment of Parkinson's disease. Nat Rev Drug Discov. 2006 Oct;5(10):845–54. doi:10.1038/nrd2087.

Shults CW, et al. Carbidopa/levodopa and selegiline do not affect platelet mitochondrial function in early Parkinsonism. Neurol. 1995 Feb;45(2):344–8. doi:10.1212/WNL.45.2.344.

Shults CW, et al. Coenzyme Q10 levels correlate with the activities of complexes I and II/III in mitochondria from parkinsonian and nonparkinsonian subjects. Ann Neurol. 1997 Aug. 42(2):261–4. doi:10.1002/ana.410420221.

Shults CW, et al. Absorption, tolerability, and effects on mitochondrial activity of oral coenzyme Q10 in parkinsonian patients. Neurology. 1998 Mar;50(3):793–5. doi:10.1212/WNL.50.3.793.

Shults CW, Haas RH, Beal MF. A possible role of coenzyme Q10 in the etiology and treatment of Parkinson's disease. Biofactors. 1999;9(2–4):267–72. doi:10.1002/biof.5520090223.

Smith TS, Parker WD, Bennell JP Jr. L-dopa increases nigral production of hydroxyl radicals *in vivo*: potential L-dopa toxicity? Neuroreportl. 1994 Apr 14;5(8):1009–11. doi:10.1097/00001756-199404000-00039.

Subramaniam SR, Chesselet MF. Mitochondrial dysfunction and oxidative stress in Parkinson's disease. Prog Neurobiol. 2013 Jul–Aug;106–107:17–32. Epub 2013 Apr 30. doi:10.1016/j.pneurobio.2013.04.004.

Thomas B, Beal MF. Mitochondrial therapies for Parkinson's disease. Mov Disord. 2010;25 Suppl 1:S155–S160. doi:10.1002/mds.22781.

Trempe JF, Fon EA. Structure and function of Parkin, PINK1, and DJ-1, the Three Musketeers of neuroprotection. Front Neurol. 2013 Apr 19;4:38. doi:10.3389/fneur.2013.00038.

Wu RM, et al. Apparent antioxidant effect of L-deprenyl on hydroxyl radical generation and nigral injury elicited by MPP+ in vivo. Eur J Pharmacol. 1993 Oct 26;243(3):241–7. doi:10.1016/0014-2999(93)90181-G.

Depression

Hroudova J, et al. Mitochondrial respiration in blood platelets of depressive patients. Mitochondrion. 2013 Nov;13(6):795–800. Epub May 17. doi:10.1016/j.mito.2013.05.005.

Lopresti AL, Hood SD, Drummond PD. A review of lifestyle factors that contribute to important pathways associated with major depression: diet, sleep and exercise. J Affect Disord. 2013 May 15;148(1):12–27. Epub Feb 14. doi:10.1016/j.jad.2013.01.014.

Morava E, Kozicz T. The economy of stress (mal)adaptation. Neurosci Biobehav Rev. 2013 May;37(4):668–80. Epub 2013 Feb 13. doi:10.1016/j.neubiorev.2013.02.005.

Seibenhener ML, et al. Behavioral effects of SQSTM1/p62 overexpression in mice: support for a mitochondrial role in depression and anxiety. Behav Brain Res. 2013 Jul 1;248:94–103. Epub Apr 13. doi:10.1016/j.bbr.2013.04.006.

Tobe EH. Mitochondrial dysfunction, oxidative stress, and major depressive disorder. Neuropsychiatr Dis Treat. 2013;9:567–73. Epub 2013 Apr 26. doi:10.2147/NDT.S44282.

Attention-Deficit/Hyperactivity Disorder: Pay Attention to the Mitochondria

Attwell D, Gibb A. Neuroenergetics and the kinetic design of excitatory synapses. Nat Rev Neurosci. 2005 Nov;6(11):841–9. doi:10.1038/nrn1784.

Barkley RA. Behavioral inhibition, sustained attention, and executive functions: constructing a unifying theory of ADHD. Psychol Bull. 1997 Jan;121(1):65–94. doi:10.1037/0033-2909 .121.1.65.

Castellanos FX, Tannock R. Neuroscience of attention-deficit/hyperactivity disorder: the search for endophenotypes. Nat Rev Neurosci. 2002 Aug;3(8):617–628. doi:10.1038/nrn896.

Charlton RA, et al. White matter damage on diffusion tensor imaging correlates with age-related cognitive decline. Neurology. 2006 Jan 24;66(2):217–22. doi:10.1212/01. wnl.0000194256.15247.83.

Chovanova Z, et al. Effect of polyphenolic extract, pycnogenol, on the level of 8-oxoguanine in children suffering from attention deficit/hyperactivity disorder. Free Radic Res. 2006 Sep;40(9):1003–10. doi:10.1080/10715760600824902.

Cotter DR, Pariante CM, Everall IP. Glial cell abnormalities in major psychiatric disorders: the evidence and implications. Brain Res Bull. 2001 Jul 15;55(5):585–95. doi:10.1016 /S0361-9230(01)00527-5.

Dienel GA. Astrocytic energetics during excitatory neurotransmission: what are contributions of glutamate oxidation and glycolysis? Neurochem Int. 2013 Oct;63(4):244–58. Epub 2013 Jul 6. doi:10.1016/j.neuint.2013.06.015.

Dvorakova M, et al. The effect of polyphenolic extract from pine bark, pycnogenol on the level of glutathione in children suffering from attention deficit hyperactivity disorder (ADHD). Redox Rep. 2006;11(4):163–72. doi:10.1179/135100006X116664.

Dvorakova M, et al. Urinary catecholamines in children with attention deficit hyperactivity disorder (ADHD): modulation by a polyphenolic extract from pine bark (pycnogenol). Nutr Neurosci. 2007 Jun–Aug;10(3–4):151–7. doi:10.1080/09513590701565443.

Ernst M, et al. Intravenous dextroamphetamine and brain glucose metabolism. Neuropsychopharmacology. 1997 Dec,17(6):391–401. doi:10.1016/S0893-133X(97)00088-2.

Fagundes AO, et al. Chronic administration of methylphenidate activates mitochondrial respiratory chain in brain of young rats. Int J Dev Neurosci. 2007 Feb;25(1):47–51. Epub 2006 Dec 22. doi:10.1016/j.ijdevneu.2006.11.001.

Gladden LB. Lactate metabolism: a new paradigm for the third millennium. J Physiol. 2004 Jul 1;558(1):5–30.

Hansson E, Ronnback L. Altered neuronal-glial signaling in glutamatergic transmission as a unifying mechanism in chronic pain and mental fatigue. Neurochem Res. 2004 May;29(5):989–96.

Hirst WD, et al. Cultured astrocytes express messenger RNA for multiple serotonin receptor subtypes, without functional coupling of 5-HT1 receptor subtypes to adenylyl cyclase. Brain Res Mol Brain Res. 1998 Oct 30;61(1–2):90–9. doi:10.1016/S0169-328X(98)00206-X.

Jessen KR. Glial cells. Int J Biochem Cell Biol. 2004 Oct;36(10):1861–7. doi:10.1016/j.biocel .2004.02.023.

Karayanidis F, et al. ERP differences in visual attention processing between attention-deficit hyperactivity disorder and control boys in the absence of performance differences. Psychophysiology. 2000 May;37(3):319–33. doi:10.1111/1469-8986.3730319.

Kasischke KA, et al. Neural activity triggers neuronal oxidative metabolism followed by astrocytic glycolysis. Science. 2004 Jul 2;305(5608):99–103. doi:10.1126/science.1096485.

Klorman R, et al. Methylphenidate speeds evaluation processes of attention deficit disorder adolescents during a continuous performance test. J Abnorm Child Psychol. 1991 Jun;19(3):263–83.

Lepine R, Parrouillet P, Camos V. What makes working memory spans so predictive of high-level cognition? Psychon Bull Rev. 2005 Feb;12(1):165–70.

Magistretti PJ, Pellerin L. Cellular mechanisms of brain energy metabolism and their relevance to functional brain imaging. Philos Trans R Soc Lond B Biol Sci. 1999 Jul 29;354(1387):1155–63. doi:10.1098/rstb.1999.0471.

Miyazaki I, et al. Direct evidence for expression of dopamine receptors in astrocytes from basal ganglia. Brain Res. 2004 Dec 10;1029(1):120–3. doi:10.1016/j.brainres.2004.09.014.

Moldrich RX, et al. Astrocyte mGlu(2/3)-mediated cAMP potentiation is calcium sensitive: studies in murine neuronal and astrocyte cultures. Neuropharmacology. 2002 Aug;43(2):189–203. doi:10.1016/S0028-3908(02)00111-9.

Ostrow LW, Sachs F. Mechanosensation and endothelin in astrocytes—hypothetical roles in CNS pathophysiology. Brain Res Brain Res Rev. 2005 Jun;48(3):488–508. doi:10.1016/j.brainresrev.2004.09.005.

Pellerin L. How astrocytes feed hungry neurons. Mol Neurobiol. 2005 Aug;32(1):59–72. doi:10.1385/MN:32:1:059.

Pellerin L, Magistretti PJ. Ampakine CX546 bolsters energetic response of astrocytes: a novel target for cognitive-enhancing drugs acting as alpha-amino-3-hydroxy5-methyl-4-isoxazolepropionic acid (AMPA) receptor modulators. J Neurochem. 2005 Feb;92(3):668–77. doi:10.1111/j.1471-4159.2004.02905.x.

Perchet C, et al. Attention shifts and anticipatory mechanisms in hyperactive children: an ERP study using the Posner paradigm. Biol Psychiatry. 2001 Jul 1;50(1):44–57. doi:10.1016/S0006-3223(00)01119-7.

Potgieter S, Vervisch J, Lagae L. Event related potentials during attention tasks in VLBW children with and without attention deficit disorder. Clin Neurophysiol. 2003 Oct;114(10):1841–9. doi:10.1016/S1388-2457(03)00198-6.

Ronnback L, Hansson E. On the potential role of glutamate transport in mental fatigue. J Neuroinflammation. 2004 Nov;1(22).

Ross BM, et al. Increased levels of ethane, a non-invasive marker of n-3 fatty acid oxidation, in breath of children with attention deficit hyperactivity disorder. Nutr Neurosci. 2003 Oct;6(5):277–81. doi:10.1080/10284150310001612203.

Sagvolden T, et al. A dynamic developmental theory of attention-deficit/hyperactivity disorder (ADHD) predominantly hyperactive/impulsive and combined subtypes. Behav Brain Sci. 2005 Jun;28(3):397–419. doi:10.1017/S0140525X05000075.

Sanchez-Abarca LI, Tabernero A, Medina JM. Oligodendrocytes use lactate as a source of energy and as a precursor of lipids. Glia. 2001 Dec;36(3):321–9. doi:10.1002/glia.1119.

Sergeant J. The cognitive-energetic model: an empirical approach to attention-deficit hyperactivity disorder. Neurosci Biobehav Rev. 2000 Jan;24(1):7–12. doi:10.1016/S0149-7634(99)00060-3.

Sergeant JA, et al. The top and the bottom of ADHD: a neuropsychological perspective. Neurosci Biobehav Rev. 2003 Nov;27(7):583–92. doi:10.1016/j.neubiorev.2003.08.004.

Smithee JA, et al. Methylphenidate does not modify the impact of response frequency or stimulus sequence on performance and event-related potentials of children with attention deficit hyperactivity disorder. J Abnorm Child Psychol. 1998 Aug;26(4):233–45.

Sonuga-Barke EJ. The dual pathway model of AD/HD: an elaboration of neuro-developmental characteristics. Neurosci Biobehav Rev. 2003 Nov;27(7):593–604. doi:10.1016/j.neubiorev.2003.08.005.

Sunohara GA, et al. Effect of methylphenidate on attention in children with attention deficit hyperactivity disorder (ADHD): ERP evidence. Neuropsychopharmacology. 1999;21:218–28. doi:10.1016/S0893-133X(99)00023-8.

Todd RD, Botteron KN. Is attention-deficit/hyperactivity disorder an energy deficiency syndrome? Biol Psychiatry. 2001 Aug 1;50(3):151–8. doi:10.1016/S0006-3223(01)01173-8.

Volkow ND, et al. Differences in regional brain metabolic responses between single and repeated doses of methylphenidate. Psychiatry Res. 1998 Jul 15;83(1):29–36. doi:10.1016 /S0925-4927(98)00025-0.

West J, et al. Response inhibition, memory and attention in boys with attention-deficit/ hyperactivity disorder. Educational Psychology. 2002;22:533–51.

Zametkin A, et al. Cerebral glucose metabolism in adults with hyperactivity of childhood onset. N Engl J Med. 1990 Nov 15;323(20):1361–6. doi:10.1056/NEJM199011153232001.

Chronic Fatigue Syndrome, Myalgic Encephalomyelitis, and Fibromyalgia

Aaron LA, Buchwald D. Chronic diffuse musculoskeletal pain, fibromyalgia and co-morbid unexplained clinical conditions. Best Pract Res Clin Rheumatol. 2003 Aug;17(4):563–74. doi:10.1016/S1521-6942(03)00033-0.

Baraniuk JN, et al. A chronic fatigue syndrome – related proteome in human cerebrospinal fluid. BMC Neurol. 2005 Dec;5:22. doi:10.1186/1471-2377-5-22.

Barnes PR, et al. Skeletal muscle bioenergetics in the chronic fatigue syndrome. J Neurol Neurosurg Psychiatry. 1993 Jun;56(6):679–83. doi:10.1136/jnnp.56.6.679.

Bengtsson A, Henriksson KG. The muscle in fibromyalgia—a review of Swedish studies. J Rheumatol Suppl. 1989 Nov;19:144–9.

Brenu EW, et al. Immunological abnormalities as potential biomarkers in chronic fatigue syndrome/myalgic encephalomyelitis. J Transl Med. 2011 May 28;9:81. doi:10.1186/1479 -5876-9-81.

Brown MM, Jason LA. Functioning in individuals with chronic fatigue syndrome: increased impairment with co-occurring multiple chemical sensitivity and fibromyalgia. Dyn Med. 2007 Jul 30;6:9. doi:10.1186/1476-5918-6-9.

Buchwald D, Garrity D. Comparison of patients with chronic fatigue syndrome, fibromyalgia, and multiple chemical sensitivities. Arch Intern Med. 1994 Sep 26;154(18):2049–53. doi:10.1001/archinte.1994.00420180053007.

Castro-Marrero J, et al. Could mitochondrial dysfunction be a differentiating marker between chronic fatigue syndrome and fibromyalgia? Antioxid Redox Signal. 2013 Nov 20;19(15):1855–60. Epub Apr 22. doi:10.1089/ars.2013.5346.

Cordero MD, et al. Coenzyme Q(10): a novel therapeutic approach for fibromyalgia? case series with 5 patients. Mitochondrion. 2011 Jul;11(4):623–5. doi:10.1016/j.mito.2011.03.122.

Cordero MD, et al. Coenzyme Q10 in salivary cells correlate with blood cells in fibromyalgia: improvement in clinical and biochemical parameter after oral treatment. Clin Biochem. 2012 Apr;45(6):509–11. doi:10.1016/j.clinbiochem.2012.02.001.

Cordero MD, et al. Can coenzyme Q10 improve clinical and molecular parameter in fibromyalgia? Antioxid Redox Signal. 2013 Oct 20;19(12):1356–61. Epub 2013 Mar 4. doi:10.1089/ars.2013.5260.

Cordero MD, et al. Is inflammation a mitochondrial dysfunction-dependent event in fibromyalgia? Antioxid Redox Signal. 2013 Mar 1;18(7):800–7. doi:10.1089/ars.2012.4892.

Devanur LD, Kerr JR. Chronic fatigue syndrome. J Clin Virol. 2006 Nov;37(3):139–50. doi:10.1016/j.jcv.2006.08.013.

Exley C, et al. A role for the body burden of aluminium in vaccine-associated macrophagic myofasciitis and chronic fatigue syndrome. Med Hypotheses. 2009 Feb;72(2):135–9. doi:10.1016/j.mehy.2008.09.040.

Jammes Y, et al. Chronic fatigue syndrome: assessment of increased oxidative stress and altered muscle excitability in response to incremental exercise. J Intern Med. 2005 Mar;257(3): 299–310. doi:10.1111/j.1365-2796.2005.01452.x.

Kennedy G, et al. Oxidative stress levels are raised in chronic fatigue syndrome and are associated with clinical symptoms. Free Radic Biol Med. 2005 Sep 1;39(5):584–9. doi:10.1016/j.freeradbiomed.2005.04.020.

Lanea RJ, et al. Heterogeneity in chronic fatigue syndrome: evidence from magnetic resonance spectroscopy of muscle. Neuromuscul Disord. 1998 May;8(3–4):204–9. doi:10.1016/S0960-8966(98)00021-2.

Maes M. Inflammatory and oxidative and nitrosative stress pathways underpinning chronic fatigue, somatization and psychosomatic symptoms. Curr Opin Psychiatry. 2009 Jan;22(1):75–83

Manuel y Keenoy B, et al. Antioxidant status and lipoprotein peroxidation in chronic fatigue syndrome. Life Sci. 2001 Mar 16;68(17):2037–49. doi:10.1016/S0024-3205(01)01001-3.

Meeus M, et al. The role of mitochondrial dysfunctions due to oxidative and nitrosative stress in the chronic pain or chronic fatigue syndromes and fibromyalgia patients: peripheral and central mechanisms as therapeutic targets? Expert Opin Ther Targets. 2013 Sep;17(9): 1081–9. Epub Jul 9. doi:10.1517/14728222.2013.818657.

Miyamae T, et al. Increased oxidative stress and coenzyme Q10 deficiency in juvenile fibromyalgia: amelioration of hypercholesterolemia and fatigue by ubiquinol-10 supplementation. Redox Rep. 2013;18(1):12–9. doi:10.1179/1351000212Y.0000000036.

Myhill S. CFS — The central cause: mitochondrial failure [Internet]. Doctor Myhill.co.uk. [Cited 2017 June 29]. Available from: http://drmyhill.co.uk/wiki/CFS_-_The_Central_Cause: _Mitochondrial_Failure.

Myhill S, Booth NE, McLaren-Howard J. Chronic fatigue syndrome and mitochondrial dysfunction. Int J Clin Exp Med. 2009;2(1):1–16.

Nancy AL, Shoenfeld Y. Chronic fatigue syndrome with autoantibodies — the result of an augmented adjuvant effect of hepatitis-B vaccine and silicone implant. Autoimmun Rev. 2008 Oct;8(1):52–5. doi:10.1016/j.autrev.2008.07.026.

Ortega-Hernandez OD, Shoenfeld Y. Infection, vaccination, and autoantibodies in chronic fatigue syndrome, cause or coincidence? Ann N Y Acad Sci. 2009 Sep;1173:600–9. doi:10.1111/j.1749-6632.2009.04799.x.

Ozgocmen S, et al. Current concepts in the pathophysiology of fibromyalgia: the potential role of oxidative stress and nitric oxide. Rheumatol Int. 2006 May;26(7):585–97. doi:10.1007 /s00296-005-0078-z.

Villanova M, et al. Mitochondrial myopathy mimicking fibromyalgia syndrome. Muscle Nerve. 1999 Feb;22(2):289–91. doi:10.1002/(SICI)1097-4598(199902)22:2<289::AID-MUS26>3.0 .CO;2-O.

Zhang C, et al. Unusual pattern of mitochondrial DNA deletions in skeletal muscle of an adult human with chronic fatigue syndrome. Hum Mol Genet. 1995;4:751–4. doi:10.1093/hmg /4.4.751.

Type 2 Diabetes

Alikhani Z, et al. Advanced glycation end products enhance expression of pro-apoptotic genes and stimulate fibroblast apoptosis through cytoplasmic and mitochondrial pathways. J Biol Chem. 2005 Apr 1;280(13):12087–95. doi:10.1074/jbc.M406313200.

Allister EM, et al. UCP2 regulates the glucagon response to fasting and starvation. Diabetes. 2013 May;62(5):1623–33. Epub 2013 Feb 22. doi:10.2337/db12-0981.

Bach D, et al. Mitofusin-2 determines mitochondrial network architecture and mitochondrial metabolism. A novel regulatory mechanism altered in obesity. J Biol Chem. 2003 May 9; 278(19):17190–7. doi:10.1074/jbc.M212754200.

Barbosa MR, et al. Hydrogen peroxide production regulates the mitochondrial function in insulin resistant muscle cells: effect of catalase overexpression. Biochim Biophys Acta. 2013 Oct;1832(10):1591–604. Epub 2013 May 2. doi:10.1016/j.bbadis.2013.04.029.

Befroy DE, et al. Impaired mitochondrial substrate oxidation in muscle of insulin-resistant offspring of type 2 diabetic patients. Diabetes. 2007 May;56(5):1376–81. Epub 2007 Feb 7. doi:10.2337/db06-0783.

Feng B, Ruiz MA, Chakrabarti S. Oxidative-stress-induced epigenetic changes in chronic diabetic complications. Can J Physiol Pharmacol. 2013 Mar;91(3):213–20. doi:10.1139/cjpp-2012-0251.

Fiorentino TV, et al. Hyperglycemia-induced oxidative stress and its role in diabetes mellitus related cardiovascular diseases. Curr Pharm Des. 2013;19(32):5695–703. Epub 2013 Feb 20. doi:10.2174/1381612811319320005.

Frohnert BI, Bernlohr DA. Protein carbonylation, mitochondrial dysfunction, and insulin resistance. Adv Nutr. 2013 Mar 1;4(2):157–63. doi:10.3945/an.112.003319.

Goodpaster BH. Mitochondrial deficiency is associated with insulin resistance. Diabetes. 2013 Apr;62(4):1032–5. doi:10.2337/db12-1612.

Graier WF, Malli R, Kostner GM. Mitochondrial protein phosphorylation: instigator or target of lipotoxicity? Trends Endocrinol Metab. 2009 May;20(4):186–93. doi:10.1016/j.tem.2009.01.004.

Hamilton JA, Kamp F. How are free fatty acids transported in membranes? Is it by proteins or by free diffusion through the lipids? Diabetes. 1999 Dec;48(12):2255–69. doi:10.2337/diabetes.48.12.2255.

Hesselink MK, Schrauwen-Hinderling V, Schrauwen P. Skeletal muscle mitochondria as a target to prevent or treat type 2 diabetes mellitus. Nat Rev Endocrinol. 2016 Nov;12(11):633–45. Epub 2016 Jul 22. doi:10.1038/nrendo.2016.104.

Hipkiss AR. Aging, proteotoxicity, mitochondria, glycation, NAD and carnosine: possible inter-relationships and resolution of the oxygen paradox. Front Aging Neurosci. 2010 Mar 18;2:10. doi:10.3389/fnagi.2010.00010.

Hipkiss AR. Mitochondrial dysfunction, proteotoxicity, and aging: causes or effects, and the possible impact of NAD+-controlled protein glycation. Adv Clin Chem. 2010;50:123–50.

Ho JK, Duclos RI Jr, Hamilton JA. Interactions of acyl carnitines with model membranes: a (13) C-NMR study. J Lipid Res. 2002 Sep;43(9):1429–39. doi:10.1194/jlr.M200137-JLR200.

Kelley DE, Mandarino LJ. Fuel selection in human skeletal muscle in insulin resistance: a reexamination. Diabetes. 2000 May;49(5):677–83. doi:10.2337/diabetes.49.5.677.

Kelley DE, Simoneau JA. Impaired free fatty acid utilization by skeletal muscle in noninsulin-dependent diabetes mellitus. J Clin Invest. 1994 Dec;94(6):2349–56. doi:10.1172/JCI117600.

Kil IS, et al. Glycation-induced inactivation of NADP(+)-dependent isocitrate dehydrogenase: implications for diabetes and aging. Free Radic Biol Med. 2004 Dec 1;37(11):1765–78.

Li JM, Shah AM. Endothelial cell superoxide generation: regulation and relevance for cardiovascular pathophysiology. Am J Physiol Regul Integr Comp Physiol. 2004 Nov; 287(5):R1014–R1030. doi:10.1152/ajpregu.00124.2004.

Lin J, et al. Transcriptional co-activator PGC-1 alpha drives the formation of slow-twitch muscle fibres. Nature. 2002 Aug 15;418(6899):797–801. doi:10.1038/nature00904.

Lindroos MM, et al. m.3243A>G mutation in mitochondrial DNA leads to decreased insulin sensitivity in skeletal muscle and to progressive {beta}-cell dysfunction. Diabetes. 2009 Mar;58(3):543–9. doi:10.2337/db08-0981.

Linnane AW, Kovalenko S, Gingold EB. The universality of bioenergetic disease. Age-associated cellular bioenergetic degradation and amelioration therapy. Ann N Y Acad Sci. 1998 Nov 20;854:202–13. doi:10.1111/j.1749-6632.1998.tb09903.x.

Maasen JA. Mitochondria, body fat and type 2 diabetes: what is the connection? Minerva Med. 2008 Jun;99(3):241–51.

Maassen JA, et al. Mitochondrial diabetes: molecular mechanisms and clinical presentation. Diabetes. 2004 Feb;53 Suppl 1:S103–S109, doi:0.2337/diabetes.53.2007.S103.

Maassen JA, et al. Mitochondrial diabetes and its lessons for common type 2 diabetes. Biochem Soc Trans. 2006;34:819–23.

Morino K, et al. Reduced mitochondrial density and increased IRS-1 serine phosphorylation in muscle of insulin-resistant offspring of type 2 diabetic parents. J Clin Invest. 2005 Dec 1;115(12):3587–93. doi:10.1172/JCI25151.

Patti ME, et al. Coordinated reduction of genes of oxidative metabolism in humans with insulin resistance and diabetes: potential role of PGC1 and NRF1. Proc Natl Acad Sci U S A. 2003 Jul 8;100(14):8466–71. Epub 2003 Jun 27. doi:10.1073/pnas.1032913100.

Petersen KF, et al. Mitochondrial dysfunction in the elderly: possible role in insulin resistance. Science. 2003 May 16;300(5622):1140–2. doi:10.1126/science.1082889.

Ritov VB, et al. Deficiency of subsarcolemmal mitochondria in obesity and type 2 diabetes. Diabetes. 2005 Jan;54(1):8–14. doi:10.2337/diabetes.54.1.8.

Rocha M, et al. Mitochondrial dysfunction and oxidative stress in insulin resistance. Curr Pharm Des. 2013;19(32):5730–41. Epub Feb 20 2013.

Rocha M, et al. Perspectives and potential applications of mitochondria-targeted antioxidants in cardiometabolic diseases and type 2 diabetes. Med Res Rev. 2014 Jan;34(1):160–89. Epub 2013 May 3. doi:10.1002/med.21285.

Rovira-Llopis S, et al. Mitochondrial dynamics in type 2 diabetes: pathophysiological implications. Redox Biology. 2017 Apr;11:637–45. doi:10.1016/j.redox.2017.01.013.

Ryu MJ et al. Crif1 deficiency reduces adipose OXPHOS capacity and triggers inflammation and insulin resistance in mice. PLoS Genet. 2013 Mar;9(3):e1003356. Epub 2013 Mar 14. doi:10.1371/journal.pgen.1003356.

Schrauwen P, et al. Uncoupling protein 3 content is decreased in skeletal muscle of patients with type 2 diabetes. Diabetes. 2001 Dec 1;50(12):2870–3. doi:10.2337/diabetes.50.12 .2870.

Schrauwen P, Hesselink MK. Oxidative capacity, lipotoxicity, and mitochondrial damage in type 2 diabetes. Diabetes. 2004 Jun;53(6):1412–7. doi:10.2337/diabetes.53.6.1412.

Short KR, et al. Decline in skeletal muscle mitochondrial function with aging in humans. Proc Natl Acad Sci U S A. 2005 Apr 12;102(15):5618–23. doi:10.1073/pnas.0501559102.

Suwa M, et al. Metformin increases the PGC-1alpha protein and oxidative enzyme activities possibly via AMPK phosphorylation in skeletal muscle in vivo. J Appl Physiol (1985). 2006 Dec;101(6):1685–92. doi:10.1152/japplphysiol.00255.2006.

Takahashi Y, et al. Hepatic failure and enhanced oxidative stress in mitochondrial diabetes. Endocr J. 2008 Jul;55(3):509–14. doi:10.1507/endocrj.K07E-091.

UK Prospective Diabetes Study Group. Intensive blood-glucose control with sulphonylureas or insulin compared with conventional treatment and risk of complications in patients with type 2 diabetes (UKPDS 33). Lancet. 1998 Sep 12;352(9131):837–53. doi:10.1016/S0140 -6736(98)07019-6.

Vanhorebeek I, et al. Tissue-specific glucose toxicity induces mitochondrial damage in a burn injury model of critical illness. Crit Care Med. 2009 Apr;37(4):1355–64. doi:10.1097/CCM .0b013e31819cec17.

Vidal-Puig AJ, et al. Energy metabolism in uncoupling protein 3 gene knockout mice. J Biol Chem. 2000 May 26;275(21):16258–66. doi:10.1074/jbc.M910179199.

Wang X et al. Protective effect of oleanolic acid against beta cell dysfunction and mitochondrial apoptosis: crucial role of ERK-NRF2 signaling pathway. J Biol Regul Homeost Agents. 2013 Jan–Mar;27(1):55–67.

Weksler-Zangen S, et al. Dietary copper supplementation restores α-cell function of Cohen diabetic rats: a link between mitochondrial function and glucose stimulated insulin secretion. Am J Physiol Endocrinol Metab. 2013 May 15;304(10):E1023–E1034. Epub 2013 Mar 19. doi:10.1152/ajpendo.00036.2013.

Winder WW, Hardie DG. AMP-activated protein kinase, a metabolic master switch: possible roles in type 2 diabetes. Am J Physiol. 1999 Jul;277(1 Pt 1):E1–E10.

Yan W, et al. Impaired mitochondrial biogenesis due to dysfunctional adiponectin-AMPKPGC-1α signaling contributing to increased vulnerability in diabetic heart. Basic Res Cardiol. 2013 May;108(3):329. Epub 2013 Mar 5. doi:10.1007/s00395-013-0329-1.

Ye J. Mechanisms of insulin resistance in obesity. Front Med. 2013 Mar;7(1):14–24. Epub 2013 Mar 9. doi:10.1007/s11684-013-0262-6.

Medication-Induced Mitochondrial Damage and Disease

Abdoli N, et al. Mechanisms of the statins' cytotoxicity in freshly isolated rat hepatocytes. J Biochem Mol Toxicol. 2013 Jun;27(6):287–94. Epub 2013 Apr 23. doi:10.1002/jbt.21485.

Anedda A, Rial E, González-Barroso MM. Metformin induces oxidative stress in white adipocytes and raises uncoupling protein 2 levels. J Endocrinol. 2008 Oct;199(1):33–40. Epub 2008 Aug 7. doi:10.1677/JOE-08-0278.

Balijepalli S, Boyd MR, Ravindranath V. Inhibition of mitochondrial complex I by haloperidol: the role of thiol oxidation. Neuropharmacology. 1999 Apr;38(4):567–77. doi:10.1016/S0028 -3908(98)00215-9.

Balijepalli S, et al. Protein thiol oxidation by haloperidol results in inhibition of mitochondrial complex I in brain regions: comparison with atypical antipsychotics. Neurochem Int. 2001, 38, 425–35. doi:10.1016/S0197-0186(00)00108-X.

Beavis AD. On the inhibition of the mitochondrial inner membrane anion uniporter by cationic amphiphiles and other drugs. J Biol Chem. 1989 Jan 25;264:1508–15.

Belenky P, Camacho D, Collins JJ. Fungicidal drugs induce a common oxidative-damage cellular death pathway. Cell Rep. 2013 Feb 21;3(2):350–8. Epub 2013 Feb 14. doi:10.1016 /j.celrep.2012.12.021.

Berson A, et al. Steatohepatitis-inducing drugs cause mitochondrial dysfunction and lipid peroxidation in rat hepatocytes. Gastroenterology. 1998 Apr;114(4):764–74. doi:10.1016 /S0016-5085(98)70590-6.

Brinkman K, et al. Mitochondrial toxicity induced by nucleoside-analogue reverse transcriptase inhibitors is a key factor in the pathogenesis of antiretroviral-therapy-related lipodystrophy. Lancet. 1999 Sep 25;354(9184):1112–5. doi:10.1016/S0140-6736(99)06102-4.

Brinkman K, Kakuda TN. Mitochondrial toxicity of nucleoside analogue reverse transcriptase inhibitors: a looming obstacle for long-term antiretroviral therapy? Curr Opin Infect Dis. 2000 Feb;13(1):5–11.

Brown SJ, Desmond PV. Hepatotoxicity of antimicrobial agents. Sem Liver Dis. 2002;22(2): 157–67. doi:10.1055/s-2002-30103.

Carvalho FS, et al. Doxorubicin-induced cardiotoxicity: from bioenergetic failure and cell death to cardiomyopathy. Med Res Rev. 2014 Jan;34(1):106–35. Epub 2013 Mar 11. doi:10.1002 /med.21280.

Chan K, et al. Drug induced mitochondrial toxicity. Expert Opin Drug Metab Toxicol. 2005 Dec;1(4):655–69. doi:10.1517/17425255.1.4.655.

Chen Y, et al. Antidiabetic drug metformin (GlucophageR) increases biogenesis of Alzheimer's amyloid peptides via up-regulating BACE1 transcription. Proc Natl Acad Sci U S A. 2009 Mar 10;106(10):3907–12. doi:10.1073/pnas.0807991106.

Chitturi SMD, George JPD. Hepatotoxicity of commonly used drugs: nonsteroidal antiinflammatory drugs, antihypertensives, antidiabetic agents, anticonvulsants, lipid lowering agents, psychotropic drugs. Semin Liver Dis. 2002;22(2):169–83. doi:10.1055/s-2002-30102.

Chrysant SG. New onset diabetes mellitus induced by statins: current evidence. Postgrad Med. 2017 May;129(4):430–5. Epub 2017 Feb 24. doi:10.1080/00325481.2017.1292107.

Cullen JM. Mechanistic classification of liver injury. Toxicol Pathol. 2005;33(1):6–8. doi:10.1080/01926230590522428.

Dong H, et al. Involvement of human cytochrome P450 2D6 in the bioactivation of acetaminophen. Drug Metab. Dispos. 2000 Dec;28(12):1397–400.

Dykens JA, Will Y. The significance of mitochondrial toxicity testing in drug development. Drug Discov Today. 2007 Sep;12(17–18):777–85. doi:10.1016/j.drudis.2007.07.013.

Ezoulin MJ, et al. Differential effect of PMS777, a new type of acetylcholinesterase inhibitor, and galanthamine on oxidative injury induced in human neuroblastoma SK-N-SH cells. Neurosci Lett. 2005 Dec 2;389(2):61–5. doi:10.1016/j.neulet.2005.07.026.

Fromenty B, Pessayre D. Impaired mitochondrial function in microvesicular steatosis effects of drugs, ethanol, hormones and cytokines. J Hepatol. 1997;26 Suppl 2:43–53. doi:10.1016/S0168-8278(97)80496-5.

Gambelli S, et al. Mitochondrial alterations in muscle biopsies of patients on statin therapy. J. Submicrosc Cytol Pathol. 2004;36(1):85–9.

Gvozdjakova A, et al. Coenzyme Q10 supplementation reduces corticosteroids dosage in patients with bronchial asthma. Biofactors. 2005;25(1–4):235–40. doi:10.1002/biof.5520250129.

Han D, et al. Regulation of drug-induced liver injury by signal transduction pathways: critical role of mitochondria. Trends Pharmacol Sci. 2013 Apr;34(4):243–53. Epub 2013 Feb 27. doi:10.1016/j.tips.2013.01.009.

Jaeschke H, Bajt ML. Intracellular signaling mechanisms of acetaminophen-induced liver cell death. Toxicol Sci. 2006 Jan;89(1):31–41. doi:10.1093/toxsci/kfi336.

Kalghatgi S, et al. Bactericidal antibiotics induce mitochondrial dysfunction and oxidative damage in mammalian cells. Sci Transl Med. 2013 Jul 3;5(192):192ra85. doi:10.1126/scitranslmed.3006055.

Lambert P, et al. Chronic lithium treatment decreases neuronal activity in the nucleus accumbens and cingulate cortex of the rat. Neuropsychopharmacology. 1999;21:229–37. doi:10.1016/S0893-133X(98)00117-1.

Lee WM. Acetaminophen and the US acute liver failure study group: lowering the risks of hepatic failure. Hepatology. 2004 Jul;40(1):6–9. doi:10.1002/hep.20293.

Levy HB, Kohlhaas HK. Considerations for supplementing with coenzyme Q10 during statin therapy. Ann Pharmacother. 2006 Feb;40(2):290–4. doi:10.1345/aph.1G409.

Mansouri A, et al. Tacrine inhibits topoisomerases and DNA synthesis to cause mitochondrial DNA depletion and apoptosis in mouse liver. Hepatology. 2003 Sep;38(3):715–25. doi:10.1053/jhep.2003.50353.

Masubuchi Y, Suda C, Horie T. Involvement of mitochondrial permeability transition in acetaminophen-induced liver injury in mice. J Hepatol. 2005 Jan;42(1):110–6. doi:10.1016/j.jhep.2004.09.015.

Maurer I, Moller HJ. Inhibition of complex I by neuroleptics in normal human brain cortex parallels the extrapyramidal toxicity of neuroleptics. Mol Cell Biochem. 1997 Sep;174(1–2):255–9.

Mikus CR, et al. Simvastatin impairs exercise training adaptations. J Am Coll Cardiol. 2013 Aug 20;62(8):709–14. Epub 2013 Apr 10. doi:10.1016/j.jacc.2013.02.074.

Modica-Napolitano JS, et al. Differential effects of typical and atypical neuroleptics on mitochondrial function in vitro. Arch Pharm Res. 2003 Nov;26(11):951–9.

Mohamed TM, Ghaffar HM, El Husseiny RM. Effects of tramadol, clonazepam, and their combination on brain mitochondrial complexes. Toxicol Ind Health. 2015 Dec;31(12): 1325–33. Epub 2013 Jul 10. doi:10.1177/0748233713491814.

Musavi S, Kakkar P. Diazepam induced early oxidative changes at the subcellular level in rat brain. Mol Cell Biochem. 1998 Jan;178(1–2):41–6.

Neustadt J, Pieczenik SR. Medication-induced mitochondrial damage and disease. Mol Nutr Food Res. 2008 Jul;52(7):780–8. doi:10.1002/mnfr.200700075.

Olsen EA, Brambrink AM. Anesthetic neurotoxicity in the newborn and infant. Curr Opin Anaesthesiol. 2013 Oct;26(5):535–42. Epub 2013 Aug 29. doi:10.1097/01.aco.0000433061 .59939.b7.

Reid AB, et al. Mechanisms of acetaminophen-induced hepatotoxicity: role of oxidative stress and mitochondrial permeability transition in freshly isolated mouse hepatocytes. J Pharmacol Exp Ther. 2005 Feb;312(2):509–16. doi:10.1124/jpet.104.075945.

Roberton AM, Ferguson LR, Cooper GJ. Biochemical evidence that high concentrations of the antidepressant amoxapine may cause inhibition of mitochondrial electron transport. Toxicol Appl Pharmacol. 1988 Mar 30;93(1):118–26. doi:10.1016/0041-008X(88)90031-2.

Shah NL, Gordon FD. N-acetylcysteine for acetaminophen overdose: when enough is enough. Hepatology. 2007 Sep;46(3):939–41.

Sirvent P, et al. Simvastatin induces impairment in skeletal muscle while heart is protected. Biochem Biophys Res Commun. 2005 Dec 23;338(3):1426–34. doi:10.1016/j.bbrc.2005.10.108.

Sirvent P, et al. Simvastatin triggers mitochondria-induced Ca2+ signaling alteration in skeletal muscle. Biochem Biophys Res Commun. 2005 Apr 15;329(3):1067–75. doi:10.1016/j.bbrc .2005.02.070.

Souza ME, et al. Effect of fluoxetine on rat liver mitochondria. Biochem Pharmacol. 1994 Aug 3;48(3):535–41. doi:10.1016/0006-2952(94)90283-6.

Vaughan RA, et al. Ubiquinol rescues simvastatin-suppression of mitochondrial content, function and metabolism: implications for statin-induced rhabdomyolysis. Eur J Pharmacol. 2013 Jul 5;711(1–3):1–9. Epub 2013 Apr 24. doi:10.1016/j.ejphar.2013.04.009.

Velho JA, et al. Statins induce calcium-dependent mitochondrial permeability transition. Toxicology. 2006 Feb;219(1–3):124–32.

Wang MY, Sadun AA. Drug-related mitochondrial optic neuropathies. J Neuroophthalmol. 2013 Jun;33(2):172–8. doi:10.1097/WNO.0b013e3182901969.

Westwood FR, et al. Statin-induced muscle necrosis in the rat: distribution, development, and fibre selectivity. Toxicol Pathol. 2005;33(2):246–57. doi:10.1080/01926230590908213.

Xia Z, et al. Changes in the generation of reactive oxygen species and in mitochondrial membrane potential during apoptosis induced by the antidepressants imipramine, clomipramine, and citalopram and the effects on these changes by Bcl-2 and BclX(L). Biochem Pharmacol. 1999 May 15;57(10):1199–208.

Xue SY, et al. Nucleoside reverse transcriptase inhibitors induce a mitophagy-associated endothelial cytotoxicity that is reversed by coenzyme Q10 cotreatment. Toxicol Sci. 2013 Aug;134(2):323–34. Epub 2013 May 2. doi:10.1093/toxsci/kft105.

Yousif W. Microscopic studies on the effect of alprazolam (Xanax) on the liver of mice. Pak J Biol Sci. 2002;5(11):1220–5. doi:10.3923/pjbs.2002.1220.1225.

Zhao C, Shichi H. Prevention of acetaminophen-induced cataract by a combination of diallyl disulfide and N-acetylcysteine. J Ocul Pharmacol Ther. 1998 Aug;14(4):345–55. doi:10.1089 /jop.1998.14.345.

Mitochondrial Disease

Bainbridge, L. Understanding and coping with mitochondrial disease. Hamilton, ON: Hamilton Health Sciences; 2010.

Bertini E, D'Amico A. Mitochondrial encephalomyopathies and related syndromes [review]. Endocr Dev. 2009;14:38–52.

Debray FG, Lambert M, Mitchell GA. Disorders of mitochondrial function. Curr Opin Pediatr. 2008 Aug;20(4):471–82. doi:10.1097/MOP.0b013e328306ebb6.

DiMauro S, Schon EA. Mitochondrial respiratory-chain diseases. N Engl J Med. 2003 Jun;348(26):2656–68. doi:10.1056/NEJMra022567.

DiMauro S, et al. Diseases of oxidative phosphorylation due to mtDNA mutations. Semin Neurol. 2001 Sep;21(3):251–60. doi:10.1055/s-2001-17942.

Finsterer J. Leigh and Leigh-like syndrome in children and adults. Pediatr Neurol. 2008 Oct;39(4):223–35. doi:10.1016/j.pediatrneurol.2008.07.013.

Folkers K, Simonsen R. Two successful double-blind trials with coenzyme Q10 (vitamin Q10) on muscular dystrophies and neurogenic atrophies. Biochim Biophys Acta. 1995 May 24;1271(1):281–6.

Goldstein AC, Bhatia P, Vento JM. Mitochondrial disease in childhood: nuclear encoded. Neurotherapeutics. 2013 Apr;10(2):212–26. Epub Mar 21 2013. doi:10.1007/s13311-013-0185-6.

Kisler JE, Whittaker RG, McFarland R. Mitochondrial diseases in childhood: a clinical approach to investigation and management. Dev Med Child Neurol. 2010 May;52(5):422–33. doi:10.1111/j.1469-8749.2009.03605.x.

Koenig MK. Presentation and diagnosis of mitochondrial disorders in children. Pediatr Neurol. 2008 May;38(5):305–13. doi:10.1016/j.pediatrneurol.2007.12.001.

Li H, et al. Comparative bioenergetic study of neuronal and muscle mitochondria during aging. Free Radic Biol Med. 2013 Oct;63:30–40. Epub Apr 30 2013. doi:10.1016/j.freeradbiomed.2013.04.030.

Lodi R, et al. Antioxidant treatment improves in vivo cardiac and skeletal muscle bioenergetics in patients with Friedreich's ataxia. Ann Neurol. 2001 May 1;49(5):590–6. doi:10.1002/ana.1001.

McFarland R, Taylor RW, Turnbull DM. A neurological perspective on mitochondrial disease. Lancet Neurol. 2010 Aug;9(8):829–840. doi:10.1016/S1474-4422(10)70116-2.

Siciliano G, et al. Functional diagnostics in mitochondrial diseases. Biosci Rep. 2007 Jun;27(1–3):53–67. doi:10.1007/s10540-007-9037-0.

Sproule DM, Kaufmann P. Mitochondrial encephalopathy, lactic acidosis, and strokelike episodes: basic concepts, clinical phenotype, and therapeutic management of MELAS syndrome. Ann N Y Acad Sci. 2008 Oct;1142:133–58. doi:10.1196/annals.1444.011.

Tarnopolsky MA, Raha S. Mitochondrial myopathies: diagnosis, exercise intolerance, and treatment options. Med Sci Sports Exerc. 2005 Dec;37(12):2086–93.

Taylor RW, Turnbull DM. Mitochondrial DNA mutations in human disease. Nat Rev Genet. 2005 May;6(5):389–402. doi:10.1038/nrg1606.

Thorburn DR. Mitochondrial disorders: prevalence, myths and advances. J Inherit Metab Dis. 2004;27(3):349–62. doi:10.1023/B:BOLI.0000031098.41409.55.

Tuppen HA, et al. Mitochondrial DNA mutations and human disease. Biochim Biophys Acta. 2010 Feb;1797(2):113–28. doi:10.1016/j.bbabio.2009.09.005.

Uitto J, Bernstein EF. Molecular mechanisms of cutaneous aging: connective tissue alterations in the dermis. J Investig Dermatol Symp Proc. 1998 Aug;3(1):41–4.

Waller JM, Maibach HI. Age and skin structure and function, a quantitative approach (II): protein, glycosaminoglycan, water, and lipid content and structure. Skin Res Technol. 2006 Aug;12(3):145–54. doi:10.1111/j.0909-752X.2006.00146.x.

Age-Related Hearing Loss

Bai U, et al. Mitochondrial DNA deletions associated with aging and possibly presbycusis: a human archival temporal bone study. Am J Otol. 1997 Jul;18(4):449–53.

Chen FQ, et al. Mitochondrial peroxiredoxin 3 regulates sensory cell survival in the cochlea. PLoS One. 2013 Apr 23;8(4):e61999. doi:10.1371/journal.pone.0061999.

Dahl HH, et al. Etiology and audiological outcomes at 3 years for 364 children in Australia. PLoS One. 2013;8(3):e59624. Epub 2013 Mar 28. doi:10.1371/journal.pone.0059624.

Ding Y, et al. The role of mitochondrial DNA mutations in hearing loss. Biochem Genet. 2013 Aug;51(7–8):588–602. Epub Apr 21 2013. doi:10.1007/s10528-013-9589-6.

Granville DJ, Gottlieb RA. Mitochondria: Regulators of cell death and survival. Scientific World Journal. 2002 Jun 11;2:1569–78. doi:10.1100/tsw.2002.809.

Han C, Someya S. Maintaining good hearing: calorie restriction, Sirt3, and glutathione. Exp Gerontol. 2013 Oct 1;48(10):1091–5. Epub 2013 Feb 20. doi:10.1016/j.exger.2013.02.014.

Johnsson LG, Hawkins JE Jr. Vascular changes in the human inner ear associated with aging. Ann Otol Rhinol Laryngol. 1972 Jun;81(3):364–76. doi:10.1177/000348947208100307.

Komlosi K, et al. Non-syndromic hearing impairment in a Hungarian family with the m.7510T>C mutation of mitochondrial tRNA(Ser(UCN)) and review of published cases. JIMD Rep. 2013;9:105–11. Epub 2012 Nov 2. doi:10.1007/8904_2012_187.

Lin FR, et al. Hearing loss and cognitive decline in older adults. JAMA Intern Med. 2013;173(4):293–9. doi:10.1001/jamainternmed.2013.1868.

Luo LF, Hou CC, Yang WX. Nuclear factors: roles related to mitochondrial deafness. Gene. 2013 May 15;520(2):79–89. Epub 2013 Mar 17. doi:10.1016/j.gene.2013.03.041.

Miller JM, Marks NJ, Goodwin PC. Laser Doppler measurements of cochlear blood flow. Hearing Res. 1983 Sep;11(3):385–94.

Seidman MD. Effects of dietary restriction and antioxidants on presbycusis. Laryngoscope. 2000 May;110(5 pt 1):727–38. doi:10.1097/00005537-200005000-00003.

Seidman MD, et al. Age related differences in cochlear microcirculation and auditory brain stem responses. Arch Otolaryngol Head Neck Surg. 1996 Nov;122(11):1221–6. doi:10.1001/archotol.1996.01890230067013.

Seidman MD, et al. Mitochondrial DNA deletions associated with aging and presbycusis. Arch Otolaryngol Head Neck Surg. 1997 Oct;123(10):1039–45.

Seidman MD, et al. Biologic activity of mitochondrial metabolites on aging and age-related hearing loss. Am J Otol. 2000 Mar;21(2):161–7.

Seidman MD, Moneysmith M. Save your hearing now. New York: Warner Books; 2006.

Semsei I, Rao G, Richardson A. Changes in the expression of superoxide dismutase and catalase as a function of age and dietary restriction. Biochem Biophys Res Commun. 1989 Oct 31;164(2):620–5. doi:10.1016/0006-291X(89)91505-2.

Wallace DC. Mitochondrial genetics: a paradigm for aging and degenerative diseases? Science. 1992 May 1;256(5057):628–32. doi:10.1126/science.1533953.

Yamasoba T, et al. Current concepts in age-related hearing loss: epidemiology and mechanistic pathways. Hear Res. 2013 Sep;303:30–8. Epub 2013 Feb 16. doi:10.1016/j.heares.2013.01.021.

Yelverton JC, et al. The clinical and audiologic features of hearing loss due to mitochondrial mutations. Otolaryngol Head Neck Surg. 2013 Jun;148(6):1017–22. Epub 2013 Mar 22. doi:10.1177/0194599813482705.

Mitochondria, Aging Skin, and Wrinkles

Balin AK, Pratt LA. Physiological consequences of human skin aging. Cutis. 1989 May;43(5):431–6.

Blatt T, et al. Stimulation of skin's energy metabolism provides multiple benefits for mature human skin. Biofactors. 2005;25(1–4):179–85. doi:10.1002/biof.5520250121.

Greco M, et al. Marked aging-related decline in efficiency of oxidative phosphorylation in human skin fibroblasts. FASEB J. 2003 Sep;17(12):1706–8. doi:10.1096/fj.02-1009fje.

Kagan J, Srivastava S. Mitochondria as a target for early detection and diagnosis of cancer. Crit Rev Clin Lab Sci. 2008;42(5–6):453–72. doi:10.1080/10408360500295477.

Kleszczynski K, Fischer TW. Melatonin and human skin aging. Dermatoendocrinol. 2012 Jul 1;4(3):245–52. doi:10.4161/derm.22344.

Kurban RS, Bhawan J. Histologic changes in skin associated with aging. J Dermatol Surg Oncol. 1990 Oct;16(10):908–14.

Navarro A, Boveris A. The mitochondrial energy transduction system and the aging process. Am J Physiol Cell Physiol. 2007 Feb;292(2):C670–C686. Epub 2006 Oct 4. doi:10.1152/ajpcell.00213.2006.

Passi S, et al. Lipophilic antioxidants in human sebum and aging. Free Radic Res. 2002 Apr;36(4):471–7.

Passi S, et al. The combined use of oral and topical lipophilic antioxidants increases their levels both in sebum and stratum corneum. Biofactors. 2003;18(1–4):289–97. doi:10.1002/biof.5520180233.

Rusciani L, et al. Low plasma coenzyme Q10 levels as an independent prognostic factor for melanoma progression. J Am Acad Dermatol. 2006 Feb;54(2):234–41. doi:10.1016/j.jaad.2005.08.031.

Treiber N, et al. The role of manganese superoxide dismutase in skin aging. Dermatoendocrinol. 2012 Jul 1;4(3):232–5. doi:10.4161/derm.21819.

Uitto J, Bernstein EF. Molecular mechanisms of cutaneous aging: connective tissue alterations in the dermis. J Investig Dermatol Symp Proc. 1998 Aug;3(1):41–4.

Waller JM, Maibach HI. Age and skin structure and function, a quantitative approach (II): protein, glycosaminoglycan, water, and lipid content and structure. Skin Res Technol. 2006 Aug;12(3):145–54. doi:10.1111/j.0909-752X.2006.00146.x.

Infertility and Mitochondria

Al Rawi S, et al. Postfertilization autophagy of sperm organelles prevents paternal mitochondrial DNA transmission. Science. 2011 Nov 25;334(6059):1144–7. Epub 2011 Oct 27. doi:10.1126/science.1211878.

Baylis F. The ethics of creating children with three genetic parents. Reprod Biomed Online. 2013 Jun; 26(6):531–4. Epub 2013 Mar 26. doi:10.1016/j.rbmo.2013.03.006.

Chappel S. The role of mitochondria from mature oocyte to viable blastocyst. Obstet Gynecol Int. 2013:1–10. Epub 2013 May 16. doi:10.1155/2013/183024.

Colagar AH, et al. T4216C mutation in NADH dehydrogenase I gene is associated with recurrent pregnancy loss. Mitochondrial DNA. 2013 Oct;24(5):610–2. Epub 2013 Mar 6. doi:10.3109/19401736.2013.772150.

Cotterill M, et al. The activity and copy number of mitochondrial DNA in ovine oocytes throughout oogenesis in vivo and during oocyte maturation in vitro. Mol Hum Reprod. 2013 Jul;19(7):444–50. Epub 2013 Mar 5. doi:10.1093/molehr/gat013.

Eichenlaub-Ritter U. Oocyte aging and its cellular basis. Int J Dev Biol. 2012;56(10–12):841–52. doi:10.1387/ijdb.120141ue.

Grindler NM, Moley KH. Maternal obesity, infertility and mitochondrial dysfunction: potential mechanisms emerging from mouse model systems. Mol Hum Reprod. 2013 Aug;19(8):486–94. Epub 2013 Apr 23. doi:10.1093/molehr/gat026.

Kang E, et al. Mitochondrial replacement in human oocytes carrying pathogenic mitochondrial DNA mutations. Nature. 2016 Dec 8;540(7632):270–5. doi:10.1038/nature20592.

Latorre-Pellicer A, et al. Mitochondrial and nuclear DNA matching shapes metabolism and healthy ageing. Nature. 2016 Jul 28;535(7613):561–5. Epub 2016 Jul 6. doi:10.1038/nature 18618.

Pang W, et al. Low expression of Mfn2 is associated with mitochondrial damage and apoptosis in the placental villi of early unexplained miscarriage. Placenta. 2013 Jul;34(7):613–8. Epub 2013 Apr 17. doi:10.1016/j.placenta.2013.03.013.

Sato M, Sato K. Degradation of paternal mitochondria by fertilization-triggered autophagy in C. elegans embryos. Science. 2011 Nov 25; 334(6059):1141–4. doi:10.1126/science .1210333.

Tillett T. Potential mechanism for PM10 effects on birth outcomes: in utero exposure linked to mitochondrial DNA damage. Environ Health Perspect. 2012 Sep;120(9):A363. doi:10.1289 /ehp.120-a363b.

Zuccotti M, Redi CA, Garagna S. Study an egg today to make an embryo tomorrow. Int J Dev Biol. 2012;56(10–12):761–4. doi:10.1387/ijdb.130027mz.

Eye-Related Diseases

Banerjee D, et al. Mitochondrial genome analysis of primary open angle glaucoma patients. PLoS One. 2013 Aug 5;8(8):e70760. doi:10.1371/journal.pone.0070760.

Blasiak J, et al. Mitochondrial and nuclear DNA damage and repair in age-related macular degeneration. Int J Mol Sci. 2013 Feb;14(2):2996–3010. Epub 2013 Jan 31. doi:10.3390 /ijms14022996.

Chen SD, Wang L, Zhang XL. Neuroprotection in glaucoma: present and future. Chin Med J (Engl). 2013 Apr;126(8):1567–77. doi:10.3760/cma.j.issn.0366-6999.20123565.

Ghiso JA, et al. Alzheimer's disease and glaucoma: mechanistic similarities and differences. J Glaucoma. 2013 Jun–Jul;22 Suppl 5:S36–S38. doi:10.1097/IJG.0b013e3182934af6.

Izzotti A, et al. Mitochondrial damage in the trabecular meshwork of patients with glaucoma. Arch Ophthalmol. 2010 Jun;128(6):724–30. doi:10.1001/archophthalmol.2010.87.

Lee V, et al. Vitamin D rejuvenates aging eyes by reducing inflammation, clearing amyloid beta and improving visual function. Neurobiol Aging. 2012 Oct;33(10):2382–9. Epub 2012 Jan 2. doi:10.1016/j.neurobiolaging.2011.

Wang MY, Sadun AA. Drug-related mitochondrial optic neuropathies. J Neuroophthalmol. 2013 Jun;33(2):172–8. doi:10.1097/WNO.0b013e3182901969.

Stem Cells Require Healthy Mitochondria

Conboy IM, Rando TA. Aging, stem cells and tissue regeneration: lessons from muscle. Cell Cycle. 2005 Mar;4(3):407–10. doi:10.4161/cc.4.3.1518.

Flynn JM, Melov S. SOD2 in mitochondrial dysfunction and neurodegeneration. Free Radic Biol Med. 2013 Sep;62:4–12. Epub May 29 2013. doi:10.1016/j.freeradbiomed.2013.05.027.

Garcia ML, Fernandez A, Solas MT. Mitochondria, motor neurons and aging. J Neurol Sci. 2013 Jul 15. Epub 2013 Apr 26. doi:10.1016/j.jns.2013.03.019.

Hosoe K, et al. Study on safety and bioavailability of ubiquinol (Kaneka QH) after single and 4-week multiple oral administration to healthy volunteers. Regul Toxicol Pharmacol. 2007 Feb;47(1):19–28. doi:10.1016/j.yrtph.2006.07.001.

Katajisto P, et al. Stem cells. Asymmetric apportioning of aged mitochondria between daughter cells is required for stemness. Science. 2015 Apr 17;348(6232):340–3. Epub 2015 Apr 2. doi:10.1126/science.1260384.

Sahin E, DePinho RA. Linking functional decline of telomeres, mitochondria and stem cells during ageing. Nature. 2010 Mar 25;464(7288):520–8. doi:0.1038/nature08982.

Cancers: Understanding the Causes Brings Us One Step Closer to Cures

Adams JS, Cory S. The Bcl-2 protein family: arbiters of cell survival. Science. 1998 Aug 28; 281(5318):1322–6.

Brown JM. Tumor microenvironment and the response to anticancer therapy. Cancer Biol Ther. 2002 Sep–Oct;1(5):453–8. doi:10.4161/cbt.1.5.157.

Bui T, Thompson CB. Cancer's sweet tooth. Cancer Cell. 2006 Jun;9(6):419–20. doi:10.1016 /j.ccr.2006.05.012.

Carracedo A, Cantley LC, Pandolfi PP. Cancer metabolism: fatty acid oxidation in the limelight. Nat Rev Cancer. 2013 Apr;13(4):227–32. Epub 2013 Feb 28. doi:10.1038/nrc3483.

Christofferson, T. Tripping over the truth: how the metabolic theory of cancer is overturning one of medicine's most entrenched paradigms. White River Junction, VT: Chelsea Green Publishing; 2017.

Dalla Via L, et al. Mitochondrial permeability transition as target of anticancer drugs. Curr Pharm Des. 2014;20(2):223–44. Epub 2013 May 16.

Davila AF, Zamorano P. Mitochondria and the evolutionary roots of cancer. Phys Biol. 2013 Apr;10(2):026008. Epub 2013 Mar 22. doi:10.1088/1478-3975/10/2/026008.

DeBerardinis RJ, et al. Beyond aerobic glycolysis: transformed cells can engage in glutamine metabolism that exceeds the requirement for protein and nucleotide synthesis. Proc Natl Acad Sci U S A. 2007 Dec 4;104(49):19345–50. doi:10.1073/pnas.0709747104.

DeBerardinis RJ, et al. The biology of cancer: metabolic reprogramming fuels cell growth and proliferation. Cell Metab. 2008 Jan;7(1):11–20. doi:10.1016/j.cmet.2007.10.002.

Fantin VR, St-Pierre J, Leder P. Attenuation of LDH-A expression uncovers a link between glycolysis, mitochondrial physiology, and tumor maintenance. Cancer Cell. 2006 Jun; 9(6):425–34. doi:10.1016/j.ccr.2006.04.023.

Gottfried E, et al. Tumor-derived lactic acid modulates dendritic cell activation and antigen expression. Blood. 2006 Mar 1;107(5):2013–21. doi:10.1182/blood-2005-05-1795.

Gottlieb E, Tomlinson IP. Mitochondrial tumour suppressors: a genetic and biochemical update. Nat Rev Cancer. 2005 Nov;5(11):857–66. doi:10.1038/nrc1737.

He X, et al. Suppression of mitochondrial complex I influences cell metastatic properties. PLoS One. 2013 Apr 22;8(4):e61677. doi:10.1371/journal.pone.0061677.

Hoang BX, et al. Restoration of cellular energetic balance with L-carnitine in the neuro-bioenergetic approach for cancer prevention and treatment. Med Hypotheses. 2007;69(2): 262–72. doi:10.1016/j.mehy.2006.11.049.

Hung WY, et al. Somatic mutations in mitochondrial genome and their potential roles in the progression of human gastric cancer. Biochim Biophys Acta. 2010 Mar;1800(3):264–70. doi:10.1016/j.bbagen.2009.06.006.

Ishikawa K, et al. ROS-generating mitochondrial DNA mutations can regulate tumor cell metastasis. Science. 2008 May 2; 320(5876):661–4. doi:10.1126/science.1156906.

Kiebish MA, et al. Cardiolipin and electron transport chain abnormalities in mouse brain tumor mitochondria: lipidomic evidence supporting the Warburg theory of cancer. J Lipid Res. 2008 Dec;49(12):2545–66. doi:10.1194/jlr.M800319-JLR200.

Kroemer G, Pouyssegur J. Tumor cell metabolism: cancer's Achilles' heel. Cancer Cell. 2008 Jun;13(6):472–82. doi:10.1016/j.ccr.2008.05.005.

Kulawiec M, Owens KM, Singh KK. Cancer cell mitochondria confer apoptosis resistance and promote metastasis. Cancer Biol Ther. 2009 Jul;8(14):1378–85.

Ladiges W, et al. A mitochondrial view of aging, reactive oxygen species and metastatic cancer. Aging Cell. 2010 Aug;9(4):462–5. doi:10.1111/j.1474-9726.2010.00579.x.

Lee HC, Chang CM, Chi CW. Somatic mutations of mitochondrial DNA in aging and cancer progression. Ageing Res Rev. 2010 Nov;9 Suppl 1:S47–S58. doi:10.1016/j.arr.2010.08.009.

Li X, et al. Targeting mitochondrial reactive oxygen species as novel therapy for inflammatory diseases and cancers. J Hematol Oncol. 2013 Feb 25;6(1):19. Epub. doi:10.1186/1756-8722 -6-19.

Lin CC, et al. Loss of the respiratory enzyme citrate synthase directly links the Warburg effect to tumor malignancy. Sci Rep. 2012;2:785. Epub 2012 Nov 8. doi:10.1038/srep00785.

Ma Y, et al. Mitochondrial dysfunction in human breast cancer cells and their transmitochondrial cybrids. Biochim Biophys Acta. 2010 Jan; 1797(1):29–37. doi:10.1016 /j.bbabio.2009.07.008.

Modica-Napolitano JS, Kulawiec M, Singh KK. Mitochondria and human cancer. Curr Mol Med. 2007 Feb;7(1):121–31. doi:10.2174/156652407779940495.

Nicolson GL, Conklin KA. Reversing mitochondrial dysfunction, fatigue and the adverse effects of chemotherapy of metastatic disease by molecular replacement therapy. Clin Exp Metastasis. 2008; 25(2):161–9. doi:10.1007/s10585-007-9129-z.

Ordys BB, et al. The role of mitochondria in glioma pathophysiology. Mol Neurobiol. 2010 Aug;42(1):64–75. doi:10.1007/s12035-010-8133-5.

Parr R, et al. Mitochondria and cancer. Biomed Res Int. 2013;2013:763703:1–2. Epub 2013 Jan 30. doi:10.1155/2013/763703.

Peck B, Ferber EC, Schulze A. Antagonism between FOXO and MYC regulates cellular powerhouse. Front Oncol. 2013 Apr 25;3:96. doi:10.3389/fonc.2013.00096.

Pelicano H, et al. Mitochondrial respiration defects in cancer cells cause activation of Akt survival pathway through a redox-mediated mechanism. J Cell Biol. 2006 Dec 18;175 (6):913–23. doi:10.1083/jcb.200512100.

Pratheeshkumar P, Thejass P, Kutan G. Diallyl disulfide induces caspase-dependent apoptosis via mitochondria-mediated intrinsic pathway in B16F-10 melanoma cells by up-regulating p53, caspase-3 and down-regulating pro-inflammatory cytokines and nuclear factor-kB-mediated Bcl-2 activation. J Environ Pathol Toxicol Oncol. 2010;29(2):113–25. doi:10.1080 /01635581.2012.721156.

Ralph SJ, et al. The causes of cancer revisited: "mitochondrial malignancy" and ROSinduced oncogenic transformation — why mitochondria are targets for cancer therapy. Mol Aspects Med. 2010 Apr;31(2):145–70. doi:10.1016/j.mam.2010.02.008.

Ramos-Montoya A, et al. Pentose phosphate cycle oxidative and nonoxidative balance: a new vulnerable target for overcoming drug resistance in cancer. Int J Cancer. 2006 Dec 15;119(12):2733–41. doi:10.1002/ijc.22227.

Ray S, Biswas S, Ray M. Similar nature of inhibition of mitochondrial respiration of heart tissue and malignant cells by methylglyoxal. A vital clue to understand the biochemical basis of malignancy. Mol Cell Biochem. 1997 Jun;171(1–2):95–103.

Shidara Y, et al. Positive contribution of pathogenic mutations in the mitochondrial genome to the promotion of cancer by prevention from apoptosis. Cancer Res. 2005 Mar 1;65(5): 1655–63. doi:10.1158/0008-5472.CAN-04-2012.

Singh KK. Mitochondrial dysfunction is a common phenotype in aging and cancer. Ann N Y Acad Sci. 2004 Jun;1019:260–4. doi:10.1196/annals.1297.043.

Sotgia F, Martinez-Outschoorn UE, Lisanti MP. Cancer metabolism: new validated targets for drug discovery. Oncotarget. 2013 Aug;4(8):1309–16. Epub 2013 Jul 22. doi:10.18632 /oncotarget.1182.

Walenta S, Mueller-Klieser WF. Lactate: mirror and motor of tumor malignancy. Semin Radiat Oncol. 2004 Jul;14(3):267–74. doi:10.1016/j.semradonc.2004.04.004.

Walenta S, et al. High lactate levels predict likelihood of metastases, tumor recurrence, and restricted patient survival in human cervical cancers. Cancer Res. 2000 Feb 15;60(4):916–21.

Wallace DC. Mitochondria and cancer: Warburg addressed. Cold Spring Harb Symp Quant Biol. 2005;70:363–74. doi:10.1101/sqb.2005.70.035.

Warburg O. On the origin of cancer cells. Science. 1956 Feb 24;123(3191):309–14. doi:10.1126/science.123.3191.309.

Wenzel U, Daniel H. Early and late apoptosis events in human transformed and nontransformed colonocytes are independent on intracellular acidification. Cell Physiol Biochem. 2004;14 (1–2):65–76. doi:10.1159/000076928.

Wenzel U, Nickel A, Daniel H. Increased carnitine-dependent fatty acid uptake into mitochondria of human colon cancer cells induces apoptosis. J Nutr. 2005 Jun;135(6):1510–4.

Wigfield SM, et al. PDK-1 regulates lactate production in hypoxia and is associated with poor prognosis in head and neck squamous cancer. Br J Cancer. 2008 Jun 17;98(12):1975–84. doi:10.1038/sj.bjc.6604356.

Aging as a Disease

Adachi K, et al. A deletion of mitochondrial DNA in murine doxorubicin-induced cardiotoxicity. Biochem Biophys Res Comm. 1993 Sep 15;195(2):945–51. doi:10.1006/bbrc.1993.2135.

Adachi K, et al. Suppression of the hydrazine-induced formation of megamitochondria in the rat liver by coenzyme Q10. Toxicol Pathol. 1995 Nov 1;23(6):667–76.

Arbustini E, et al. Mitochondrial DNA mutations and mitochondrial abnormalities in dilated cardiomyopathy. Am J Pathol. 1998 Nov;153(5):1501–10. doi:10.1016/S0002-9440 (10)65738-0.

Cellular nutrition for vitality and longevity. Life Extension [internet]. 2000 April [cited 2017 Aug];24–28. Available from: http://www.lifeextension.com/magazine/2000/4/cover2/page-01.

DiMauro S, et al. Mitochondria in neuromuscular disorders. Biochim Biophys Acta. 1998 Aug 10;1366(1–2):199–210. doi:10.1016/S0005-2728(98)00113-3.

Esposito LA, et al. Mitochondrial disease in mouse results in increased oxidative stress. Proc Natl Acad Sci U S A. 1999 Apr 27;96(9):4820–5.

Fontaine E, Ichas F, Bernardi P. A ubiquinone-binding site regulates the mitochondrial permeability transition pore. J Biol Chem. 1998;273:25734–40.

Fontaine E, et al. Regulation of the permeability transition pore in skeletal muscle mitochondria. Modulation by electron flow through the respiratory chain complex i. J Biol Chem. 1998 May 15;273(20):12662–8.

Geromel V, et al. The consequences of a mild respiratory chain deficiency on substrate competitive oxidation in human mitochondria. Biochem Biophys Res Comm. 1997 Aug;236:643–6.

Karbowski M, et al. Free radical-induced megamitochondria formation and apoptosis. Free Radic Biol Med. 1999 Feb;26(3–4):396–409. doi:10.1016/S0891-5849(98)00209-3.

Kopsidas G, et al. An age-associated correlation between cellular bioenergy decline and mtDNA rearrangements in human skeletal muscle. Mutat Res. 1998 Oct 12;421(1):27–36. doi:10.1016/S0027-5107(98)00150-X.

Kovalenko SA, et al. Tissue-specific distribution of multiple mitochondrial DNA rearrangements during human aging. Ann N Y Acad Sci. 1998 Nov 20;854:171–81.

Ku HH, Brunk UT, Sohal RS. Relationship between mitochondrial superoxide and hydrogen peroxide production and longevity of mammalian species. Free Radic Biol Med. 1993 Dec;15(6):621–7.

Lass A, Agarwal S, Sohal RS. Mitochondrial ubiquinone homologues, superoxide radical generation, and longevity in different mammalian species. J Biol Chem. 1997 Aug 1;272:19199–204. doi:10.1074/jbc.272.31.19199.

Lass A, Sohal RS. Comparisons of coenzyme Q bound to mitochondrial membrane proteins among different mammalian species. Free Radic Biol Med. 1999;27(1–2):220–6.

Linnane AW, et al. Mitochondrial DNA mutations as an important contributor to aging and degenerative diseases. Lancet. 1989 Mar 25;1(8639):642–5. doi:10.1016/S0140-6736(89)92145-4.

Linnane AW, et al. The universality of bioenergetic disease and amelioration with redox therapy. Biochim Biophys Acta. 1995 May 24;1271(1):191–4. doi:10.1016/0925-4439(95)00027-2.

Linnane AW, Kovalenko S, Gingold EB. The universality of bioenergetic disease. Age-associated cellular bioenergetic degradation and amelioration therapy. Ann N Y Acad Sci. 1998 Nov 20;854:202–13. doi:10.1111/j.1749-6632.1998.tb09903.x.

Martinucci S, et al. Ca2+-reversible inhibition of the mitochondrial megachannel by ubiquinone analogues. FEBS Lett. 2000 Sep;480:89–94. doi:10.1016/S0014-5793(00)01911-6.

Michikawa Y, et al. Aging-dependent large accumulation of point mutations in the human mtDNA control region for replication. Science. 1999 Oct 22;286(5440):774–9. doi:10.1126/science.286.5440.774.

Ozawa T. Genetic and functional changes in mitochondria associated with aging. Physiol Rev. 1997 Apr;77(2):425–64.

Richter C, et al. Control of apoptosis by the cellular ATP level. 1996 Jan 8;FEBS Lett 378(2): 107–10. doi:10.1016/0014-5793(95)01431-4.

Rosenfeldt FL, et al. Coenzyme Q10 in vitro normalizes impaired post-ischemic contractile recovery of aged human myocardium. Fifth China International Congress on TCVS; 2000 September; Beijing, China.

Rosenfeldt FL, et al. Response of the human myocardium to hypoxia and ischemia declines with age: correlations with increased mitochondrial DNA deletions. Ann N Y Acad Sci. 1998 Nov;854:489–90. doi:10.1111/j.1749-6632.1998.tb09938.x.

Rowland MA, et al. Coenzyme Q10 treatment improves the tolerance of the senescent myocardium to pacing stress in the rat. Cardiovasc Res. 1998 Oct;40(1):165–73.

Sohal RS, Sohal BH, Orr WC. Mitochondrial superoxide and hydrogen peroxide generation, protein oxidative damage, and longevity in different species of flies. Free Radic Biol Med. 1995 Oct;19(4):499–504. doi:10.1016/0891-5849(95)00037-X.

Susin SA, et al. Mitochondria as regulators of apoptosis: doubt no more. Biochim Biophys Acta. 1998 Aug 10;1366(1–2):151–65. doi:10.1016/S0005-2728(98)00110-8.

Turker MS. Somatic cell mutations: can they provide a link between aging and cancer? Mech Aging Dev. 2000 Aug 15;117(1–3):1–19. doi:10.1016/S0047-6374(00)00133-0.

Wallace DC. Mitochondrial diseases in man and mouse. Science. 1999 Mar 5;283(5407): 1482–8. doi:10.1126/science.283.5407.1482.

Wallace DC, et al. Mitochondrial DNA mutations in human degenerative diseases and aging. Biochim Biophys Acta. 1995 May 24;1271(1):141–51. doi:10.1016/0925-4439(95)00021-U.

Walter L, et al. Three classes of ubiquinone analogs regulate the mitochondrial permeability transition pore through a common site. J Biol Chem. 2000 July 10;275:29521–7. doi:10.1074/jbc.M004128200.

Wei YH. Oxidative stress and mitochondrial DNA mutations in human aging. Proc Soc Exp Biol Med. 1998 Jan;217(1):53–63.

Wei YH, Kao SH, Lee HC. Simultaneous increase of mitochondrial DNA deletions and lipid peroxidation in human aging. Ann N Y Acad Sci. 1996 Jun 15;786:24–43. doi:10.1111/j.1749-6632.1996.tb39049.x.

Wolvetang EJ, et al. Mitochondrial respiratory chain inhibitors induce apoptosis. 1994 Feb 14;339(1–2):40–4. doi:10.1016/0014-5793(94)80380-3.

Zhang C, et al. Varied prevalence of age-associated mitochondrial DNA deletions in different species and tissues: a comparison between human and rat. Biochem Biophys Res Comm. 1997 Jan;230(3):630–5. doi:10.1006/bbrc.1996.6020.

Chapter Three

Ames BN, Atamna H, Killilea DW. Mineral and vitamin deficiencies can accelerate the mitochondrial decay of aging. Mol Aspects Med. 2005 Aug–Oct;26(4–5):363–78. doi:10.1016/j.mam.2005.07.007.

Aw TY, Jones DP. Nutrient supply and mitochondrial function. Annu Rev Nutr. 1989 Jul; 9:229–51. doi:10.1146/annurev.nu.09.070189.001305.

Williams KL, et al. Differing effects of metformin on glycemic control by race-ethnicity. J Clin Endocrinol Metab. 2014 Sep;99(9):3160–8. Epub 2014 June 12. doi:10.1210/jc.2014-1539.

D-Ribose

Andreoli SP. Mechanisms of endothelial cell ATP depletion after oxidant injury. Pediatr Res. 1989 Jan;25(1):97–101. doi:10.1203/00006450-198901000-00021.

Asimakis G, et al. Postischemic recovery of mitochondrial adenine nucleotides in the heart. Circulation. 1992 Jul;85(6):2212–20.

Baldwin D, et al. Myocardial glucose metabolism and ATP levels are decreased two days after global ischemia. J Surg Res. 1996 Jun;63(1):35–8. doi:10.1006/jsre.1996.0218.

Befera N, et al. Ribose treatment preserves function of the remote myocardium after myocardial infarction. J Surg Res. 2007 Feb;137(2):156. doi:10.1016/j.jss.2006.12.022.

Bengtsson A, Heriksson KG, Larsson J. Reduced high-energy phosphate levels in the painful muscles of patients with primary fibromyalgia. Arth Rheum. 1986 Jul;29(7):817–21. doi:10.1002/art.1780290701.

Bengtsson A, Henriksson KG. The muscle in fibromyalgia—a review of Swedish studies. J Rheumatol Suppl. 1989 Nov;19:144–9.

Brault JJ, Terjung RL. Purine salvage to adenine nucleotides in different skeletal muscle fiber types. J Appl Physiol. 2001;91:231–8.

Chatham JC, et al. Studies of the protective effect of ribose in myocardial ischaemia by using 31P-nuclear magnetic resonance spectroscopy. Biochem Soc Proc. 1985 Oct;13(5):885–8. doi:10.1042/bst0130885.

Clay MA, et al. Chronic alcoholic cardiomyopathy. Protection of the isolated ischemic working heart by ribose. Biochem Int. 1988 Nov;17(5):791–800.

Dodd SL, et al. The role of ribose in human skeletal muscle metabolism. Med Hypotheses. 2004;62(5):819–24. doi:10.1016/j.mehy.2003.10.026.

Dow J, et al. Adenine nucleotide synthesis de novo in mature rat cardiac myocytes. Biochim Biophys Acta. 1985 Nov 20;847(2):223–7. doi:10.1016/0167-4889(85)90024-2.

Ellison GM, et al. Physiological cardiac remodelling in response to endurance exercise training: cellular and molecular mechanisms. Heart (British Cardiac Society). 2012 Jan;98(1):5–10.

Enzig S, et al. Myocardial ATP repletion with ribose infusion. Pediatr Res. 1985;19:127A.

Gebhart B, Jorgenson JA. Benefit of ribose in a patient with fibromyalgia. Pharmacotherapy. 2004 Nov;24(11):1646–8. doi:10.1592/phco.24.16.1646.50957.

Gross M, Kormann B, Zollner N. Ribose administration during exercise: effects on substrates and products of energy metabolism in healthy subjects and a patient with myoadenylate deaminase deficiency. Klin Wochenschr. 1991;69(4):151–5.

Harmsen E, et al. Enhanced ATP and GTP synthesis from hypoxanthine or inosine after myocardial ischemia. Am J Physiol. 1984 Jan;246(1 Pt 2):H37–H43.

Hass GS, et al. Reduction of postischemic myocardial dysfunction by substrate repletion during reperfusion. Circulation. 1984 Sep;70(3 Pt 2):165–74.

Hellsten Y, Skadhauge L, Bangsbo J. Effect of ribose supplementation on resynthesis of nucleotides after intense intermittent training in humans. Am J Physiol. 2004 Jan 1;286(1): R182–R188. doi:10.1152/ajpregu.00286.2003.

Ibel H, Zimmer HG. Metabolic recovery following temporary regional myocardial ischemia in the rat. J Mol Cell Cardiol. 1986;18(Suppl 4):61–5.

Ingwall JS, Weiss RG. Is the failing heart energy starved? On using chemical energy to support cardiac function. Circ Res. 2004 Jul 23;95(2):135–45. doi:10.1161/01.RES.0000137170.41939.d9.

LaNoue KF, Watts JA, Koch CD. Adenine nucleotide transport during cardiac ischemia. Am J Physiol. 1981 Nov;241(5):H663–H671.

Lund N, Bengtsson A, Thorborg P. Muscle tissue oxygen in primary fibromyalgia. Scan J Rheumatol. 1986;15(2):165–73. doi:10.3109/03009748609102084.

Mahoney JR Jr. Recovery of postischemic myocardial ATP levels and hexosemonophosphate shunt activity. Med Hypoth. 1990 Jan;31(1):21–3. doi:10.1016/0306-9877(90)90047-I.

Maron BJ, Pelliccia A. The heart of trained athletes: cardiac remodeling and the risks of sports, including sudden death. Circulation. 2006 Oct 10;114(15):1633–44. doi:10.1161/CIRCULATIONAHA.106.613562.

Muller C, et al. Effect of ribose on cardiac adenine nucleotides in a donor model for heart transplantation. Eur J Med Res. 1998 Dec 16;3(12):554–8.

Omran H, et al. D-ribose improves diastolic function and quality of life in congestive heart failure patients: a prospective feasibility study. Eur J Heart Fail. 2003 Oct;5(5):615–9. doi:10.1016/S1388-9842(03)00060-6.

Omran H, et al. D-ribose aids congestive heart failure patients. Exp Clin Cardiol. 2004 Summer;9(2):117–8.

Pauly DF, Johnson C, St Cyr JA. The benefits of ribose in cardiovascular disease. Med Hypotheses. 2003 Feb;60(2):149–51.

Pauly DF, Pepine CJ. D-ribose as a supplement for cardiac energy metabolism. J Cardiovasc Pharmacol Ther. 2000 Oct;5(4):249–58. doi:10.1054/JCPT.2000.18011.

Pauly DF, Pepine CJ. Ischemic heart disease: metabolic approaches to management. Clin Cardiol. 2004;27(8):439–4l. doi:10.1002/clc.4960270802.

Pelliccia A, Di Paolo FM, Maron BJ. The athlete's heart: remodeling, electrocardiogram and preparticipation screening. Cardiol Rev. 2002 Mar–Apr;10(2):85–90.

Perkowski D, et al. D-ribose improves cardiac indices in patients undergoing "off" pump coronary arterial revascularization. J Surg Res. 2007;137(2)295.

Pliml W, et al. Effects of ribose on exercise-induced ischaemia in stable coronary artery disease. Lancet. 1992 Aug 29;340(8818):507–10. doi:10.1016/0140-6736(92)91709-H.

Pouleur H. Diastolic dysfunction and myocardial energetics. Eur Heart J. 1990 May;11(Suppl C):30–4. doi:10.1093/eurheartj/11.suppl_C.30.

Rich BS, Havens SA. The athletic heart syndrome. Curr Sports Med Rep. 2004 Mar;3(2):84–8.

Sami H, Bittar N. The effect of ribose administration on contractile recovery following brief periods of ischemia. Anesthesiology. 1987;67(3A):A74.

Schachter CL, et al. Effects of short versus long bouts of aerobic exercise in sedentary women with fibromyalgia: a randomized controlled trial. Phys Ther. 2003 Apr;83(4):340–58.

Sinatra ST. The Sinatra solution: metabolic cardiology. Laguna Beach, CA: Basic Health Publications, Inc; 2011.

Taegtmeyer H. Metabolism—the lost child of cardiology. J Am Coll Cardiol. 2000;36(4):1386–8.

Taegtmeyer H, et al. Energy metabolism in reperfused heart muscle: Metabolic correlates to return of function. J Am Coll Cardiol. 1985 Oct;6(4):864–70. doi:10.1016/S0735-1097 (85)80496-4.

Taegtmeyer H, King LM, Jones BE. Energy substrate metabolism, myocardial ischemia, and targets for pharmacotherapy. Am J Cardiol. 1998 Sep 3;82(5A):54K–60K. doi:10.1016 /S0002-9149(98)00538-4.

Teitelbaum JE, Johnson C, St Cyr J. The use of D-ribose in chronic fatigue syndrome and fibromyalgia: a pilot study. J Altern Complement Med. 2006 Nov;12(9)857–62. doi:10.1089 /acm.2006.12.857.

Tullson PC, Terjung RL. Adenine nucleotide synthesis in exercising and endurance-trained skeletal muscle. Am J Physiol. 1991 Aug;261:C342–C347.

Van Gammeren D, Falk D, Antonio J. The effects of four weeks of ribose supplementation on body composition and exercise performance in healthy, young male recreational bodybuilders: a double-blind, placebo-controlled trial. Curr Ther Res. 2002 Aug;63(8): 486–95. doi:10.1016/S0011-393X(02)80054-6.

Wilson R, MacCarter D, St. Cyr J. D-ribose enhances the identification of hibernating myocardium. Heart Drug. 2003;3:61–2. doi:10.1159/000070908.

Zarzeczny R, et al. Influence of ribose on adenine salvage after intense muscle contractions. J Appl Physiol. 2001;91:1775–81.

Zimmer HG. Restitution of myocardial adenine nucleotides: acceleration by administration of ribose. J Physiol (Paris). 1980;76(7):769–75.

Zimmer HG. Significance of the 5-phosphoribosyl-1-pyrophosphate pool for cardiac purine and pyrimidine nucleotide synthesis: studies with ribose, adenine, inosine, and orotic acid in rats. Cardiovasc Drug Ther. 1998 Apr;12(Suppl 2):179–87.

Zimmer HG, et al. Ribose intervention in the cardiac pentose phosphate pathway is not species-specific. Science. 1984 Feb 17;223(4637):712–4. doi:10.1126/science.6420889.

Zimmer HG, Ibel H. Effects of ribose on cardiac metabolism and function in isoproterenol-treated rats. Am J Physiol. 1983 Nov;245:H880–H886.

Pyrroloquinoline Quinone (PQQ)

Aizenman E, et al. Interaction of the putative essential nutrient pyrroloquinoline quinone with the N-methyl-daspartate receptor redox modulatory site. J Neurosci. 1992 Jun;12(6):2362–9.

Aizenman E, et al. Further evidence that pyrroloquinoline quinone interacts with the N-methyl—aspartate receptor redox site in rat cortical neurons *in vitro*. Neurosci Lett. 1994 Feb 28;168(1–2):189–92. doi:10.1016/0304-3940(94)90447-2.

Bauerly KA, et al. Pyrroloquinoline quinone nutritional status alters lysine metabolism and modulates mitochondrial DNA content in the mouse and rat. Biochim Biophys Acta. 2006 Nov;1760(11):1741–8. doi:10.1016/j.bbagen.2006.07.009.

Chowanadisai W, et al. Pyrroloquinoline quinone (PQQ) stimulates mitochondrial biogenesis. FASEB J. 2007 Apr;21:854. doi:10.1074/jbc.M109.030130.

Chowanadisai W, et al. Pyrroloquinoline quinone stimulates mitochondrial biogenesis through cAMP response element-binding protein phosphorylation and increased PGC-1α expression. J Biol Chem. 2010 Jan 1;285(1):142–52. doi:10.1074/jbc.M109.030130.

Debray FG, Lambert M, Mitchell GA. Disorders of mitochondrial function. Curr Opin Pediatr. 2008 Aug;20(4):471–82. doi:10.1097/MOP.0b013e328306ebb6.

Felton LM, Anthony C. Biochemistry: role of PQQ as a mammalian enzyme cofactor? Nature. 2005 Feb 3;433(7025):E10;discussion E11–E12. doi:10.1038/nature03322.

Harris CB, et al. Dietary pyrroloquinoline quinone (PQQ) alters indicators of inflammation and mitochondrial-related metabolism in human subjects. J Nutr Biochem. 2013 Dec;24(12):2076–84. doi:10.1016/j.jnutbio.2013.07.008.

Hirakawa A, et al. Pyrroloquinoline quinone attenuates iNOS gene expression in the injured spinal cord. Biochem Biophys Res Commun. 2009 Jan 9;378(2):308–12. doi:10.1016/j.bbrc.2008.11.045.

Jensen FE, et al. The putative essential nutrient pyrroloquinoline quinone is neuroprotective in a rodent model of hypoxic/ischemic brain injury. Neuroscience. 1994 Sep;62(2):399–406. doi:10.1016/0306-4522(94)90375-1.

Kasahara T, Kato T. Nutritional biochemistry: a new redox-cofactor vitamin for mammals. Nature. 2003 Apr 24;422:832. doi:10.1038/422832a.

Kumazawa T, Seno H, Suzuki O. Failure to verify high levels of pyrroloquinoline quinone in eggs and skim milk. Biochem Biophys Res Commun. 1993 May 28;193(1):1–5. doi:10.1006/bbrc.1993.1581.

Kumazawa T, et al. Levels of pyrroloquinoline quinone in various foods. Biochem J. 1995;307:331–3. doi:10.1042/bj3070331.

Kumazawa T, et al. Activation of ras signaling pathways by pyrroloquinoline quinone in NIH3T3 mouse fibroblasts. Int J Mol Med. 2007 May;19(5):765–70. doi:10.3892/ijmm.19.5.765.

Li HH, et al. Pyrroloquinoline quinone enhances regeneration of transected sciatic nerve in rats. Chin J Traumatol. 2005 Aug;8(4):225–9.

Magnusson OT, et al. Quinone biogenesis: structure and mechanism of PqqC, the final catalyst in the production of pyrroloquinoline quinone. Proc Natl Acad Sci U S A. 2004 May 25;101(21):7913–8. doi:10.1073/pnas.0402640101.

Magnusson OT, et al. Pyrroloquinoline quinone biogenesis: characterization of PqqC and its H84N and H84A active site variants. Biochemistry. 2007;46(24):7174–86. doi:10.1021/bi700162n.

Matsushita K, et al. *Escherichia coli* is unable to produce pyrroloquinoline quinone (PQQ). Microbiology. 1997;143:3149–56. doi:10.1099/00221287-143-10-3149.

Mitchell AE, et al. Characterization of pyrroloquinoline quinone amino acid derivatives by electrospray ionization mass spectrometry and detection in human milk. Anal Biochem. 1999 May 1;269(2):317–25. doi:10.1006/abio.1999.4039.

Muoio DM, Koves TR. Skeletal muscle adaptation to fatty acid depends on coordinated actions of the PPARs and PGC-1alpha: implications for metabolic disease. Appl Physiol Nutr Metab. 2007 Oct;32(5):874–83. doi:10.1139/H07-083.

Murase K, et al. Stimulation of nerve growth factor synthesis/secretion in mouse astroglial cells by coenzymes. Biochem Mol Biol Int. 1993 Jul;30(4):615–21.

Nunome K, et al. Pyrroloquinoline quinone prevents oxidative stress-induced neuronal death probably through changes in oxidative status of DJ-1. Biol Pharm Bull. 2008 Jul;31(7):1321–6. doi:10.1248/bpb.31.1321.

Ohwada K, et al. Pyrroloquinoline quinone (PQQ) prevents cognitive deficit caused by oxidative stress in rats. J Clin Biochem Nutr. 2008 Jan;42(1):29–34. doi:10.3164/jcbn.2008005.

Ouchi A, et al. Kinetic study of the antioxidant activity of pyrroloquinolinequinol (PQQH(2), a reduced form of pyrroloquinolinequinone) in micellar solution. J Agric Food Chem. 2009;57(2):450–6. doi:10.1021/jf802197d.

Puehringer S, Metlitzky M, Schwarzenbacher R. The pyrroloquinoline quinone biosynthesis pathway revisited: a structural approach. BMC Biochem. 2008 Mar 27;9:8. doi:10.1186/1471-2091-9-8.

Puigserver P. Tissue-specific regulation of metabolic pathways through the transcriptional coactivator PGC1-alpha. Int J Obes (Lond). 2005 Mar;29:S5–S9. doi:10.1038/sj.ijo.0802905.

Rucker R, Chowanadisai W, Nakano M. Potential physiological importance of pyrroloquinoline quinone. Altern Med Rev. 2009 Sep;14(3):268–77.

Rucker R, et al. Biochemistry: is pyrroloquinoline quinone a vitamin? Nature. 2005 Feb 3;433(7025):E10–E11;discussion E11–E12. doi:10.1038/nature03323.

Sanchez RM, et al. Novel role for the NMDA receptor redox modulatory site in the pathophysiology of seizures. J Neurosci. 2000 Mar 15;20(6):2409–17.

Sato K, Toriyama M. Effect of pyrroloquinoline quinone (PQQ) on melanogenic protein expression in murine B16 melanoma. J Dermatol Sci. 2009 Feb;53(2):140–5. doi:10.1016 /j.jdermsci.2008.08.017.

Scanlon JM, Aizenman E, Reynolds IJ. Effects of pyrroloquinoline quinone on glutamate-induced production of reactive oxygen species in neurons. Eur J Pharmacol. 1997 May 12;326(1):67–74. doi:10.1016/S0014-2999(97)00137-4.

Steinberg FM, Gershwin ME, Rucker RB. Dietary pyrroloquinoline quinone: growth and immune response in BALB/c mice. J Nutr. 1994 May;124(5):744–53.

Steinberg F, et al. Pyrroloquinoline quinone improves growth and reproductive performance in mice fed chemically defined diets. Exp Biol Med (Maywood). 2003 Feb;228(2):160–6. doi:10.1177/153537020322800205.

Stites TE, Mitchell AE, Rucker RB. Physiological importance of quinoenzymes and the O-quinone family of cofactors. J Nutr. 2000 Apr;130(4):719–27.

Stites T, et al. Pyrroloquinoline quinone modulates mitochondrial quantity and function in mice. J Nutr. 2006 Feb;136(2):390–6.

Tao R, et al. Pyrroloquinoline quinone preserves mitochondrial function and prevents oxidative injury in adult rat cardiac myocytes. Biochem Biophys Res Commun. 2007 Nov 16;363(2):257–62. doi:10.1016/j.bbrc.2007.08.041.

Yamaguchi K, et al. Stimulation of nerve growth factor production by pyrroloquinoline quinone and its derivatives in vitro and in vivo. Biosci Biotechnol Biochem. 1993 Jul;57(7):1231–3. doi:10.1271/bbb.57.1231.

Zhang P, et al. Protection of pyrroloquinoline quinone against methylmercury-induced neurotoxicity via reducing oxidative stress. Free Radic Res. 2009 Mar;43(3):224–33. doi:10.1080/10715760802677348.

Zhang Y, Feustel PJ, Kimelberg HK. Neuroprotection by pyrroloquinoline quinone (PQQ) in reversible middle cerebral artery occlusion in the adult rat. Brain Res. 2006 Jun 13;1094(1): 200–6. doi:10.1016/j.brainres.2006.03.111.

Zhang Y, Rosenberg PA. The essential nutrient pyrroloquinoline quinone may act as a neuroprotectant by suppressing peroxynitrite formation. Eur J Neurosci. 2002 Sep;16(6): 1015–24. doi:10.1046/j.1460-9568.2002.02169.x.

Zhu BQ, et al. Pyrroloquinoline quinone (PQQ) decreases myocardial infarct size and improves cardiac function in rat models of ischemia and ischemia/reperfusion. Cardiovasc Drugs Ther. 2004 Nov;18(6):421–31. doi:10.1007/s10557-004-6219-x.

Zhu BQ, et al. Comparison of pyrroloquinoline quinone and/or metoprolol on myocardial infarct size and mitochondrial damage in a rat model of ischemia/reperfusion injury. J Cardiovasc Pharmacol Ther. 2006 Jun;11(2):119–28. doi:10.1177/1074248406288757.

Dark Chocolate

Al-Safi SA, et al. Dark chocolate and blood pressure: a novel study from Jordan. Curr Drug Deliv. 2011 Nov;8(6):595–9. doi:10.2174/156720111797635496.

Buitrago-Lopez A, et al. Chocolate consumption and cardiometabolic disorders: systematic review and meta-analysis. BMJ. 2011 Aug 26;343:d4488. doi:10.1136/bmj.d4488.

Ellinger S, et al. Epicatechin ingested via cocoa products reduces blood pressure in humans: a nonlinear regression model with a Bayesian approach. Am J Clin Nutr. 2012 Jun;95(6): 1365–77. Epub 2012 May 2. doi:10.3945/ajcn.111.029330.

Golomb BA, Koperski S, White HL. Association between more frequent chocolate consumption and lower body mass index. Arch Intern Med. 2012 Mar 26;172(6):519–21. doi:10.1001 /archinternmed.2011.2100.

Messerli FH. Chocolate consumption, cognitive function, and Nobel laureates. N Engl J Med. 2012 Oct 18;367(16):1562–4. Epub 2012 Oct 10. doi:10.1056/NEJMon1211064.

Nehlig A. The neuroprotective effects of cocoa flavanol and its influence on cognitive performance. Br J Clin Pharmacol. 2013 Mar;75(3):716–27. doi:10.1111/j.1365-2125.2012.04378.x.

Nogueira L, et al. (-)-Epicatechin enhances fatigue resistance and oxidative capacity in mouse muscle. J Physiol. 2011 Sep 15;589(Pt 18):4615–31. Epub 2011 Jul 25. doi:10.1113/ jphysiol.2011.209924.

Persson IA, et al. Effects of cocoa extract and dark chocolate on angiotensin-converting enzyme and nitric oxide in human endothelial cells and healthy volunteers—a nutrigenomics perspective. J Cardiovasc Pharmacol. 2011 Jan;57(1):44–50. doi:10.1097/FJC.0b013 e3181fe62e3.

Sathyapalan T, et al. High cocoa polyphenol rich chocolate may reduce the burden of the symptoms in chronic fatigue syndrome. Nutr J. 2010 Nov 22;9:55. doi:10.1186/1475-2891-9-55.

Coenzyme Q10

Cooper JM, et al. Coenzyme Q10 and vitamin E deficiency in Friedreich's ataxia: predictor of efficacy of vitamin E and coenzyme Q10 therapy. Eur J Neurol. 2008 Dec;15(12):1371–9. doi:10.1111/j.1468-1331.2008.02318.x.

Crane FL, Low H, Sun IL. Evidence for a relation between plasma membrane coenzyme Q and autism. Front Biosci (Elite Ed). 2013 Jun 1;5:1011–6.

Del Pozo-Cruz J, et al. Relationship between functional capacity and body mass index with plasma coenzyme Q10 and oxidative damage in community-dwelling elderly-people. Exp Gerontol. 2014 Apr;52:46–54. Epub 2014 Feb 7.

Duberley KE, et al. Effect of coenzyme Q10 supplementation on mitochondrial electron transport chain activity and mitochondrial oxidative stress in coenzyme Q10 deficient human neuronal cells. Int J Biochem Cell Biol. 2014 May;50:60–3. Epub 2014 Feb 15. doi:10.1016/j.biocel.2014.02.003.

Liang JM, et al. Role of mitochondrial function in the protective effects of ischaemic postconditioning on ischaemia/reperfusion cerebral damage. J Int Med Res. 2013 Jun1;41(3):618–27. Epub2013 Apr 4. doi:10.1177/0300060513476587.

Langsjoen PH, Langsjoen AM. Supplemental ubiquinol in patients with advanced congestive heart failure. Biofactors. 2008;32(1–4):119–28. doi:10.1002/biof.5520320114.

Lass A, Sohal RS. Comparisons of coenzyme Q bound to mitochondrial membrane proteins among different mammalian species. Free Radic Biol Med. 1999 Jul;27(1–2):220–6. doi:10.1016/S0891-5849(99)00085-4.

Mancuso M, et al. Coenzyme Q10 in neuromuscular and neurodegenerative disorders. Curr Drug Targets. 2010 Jan;11(1):111–21. doi:10.2174/138945010790031018.

Matthews RT, et al. Coenzyme Q10 administration increases brain mitochondrial concentrations and exerts neuroprotective effects. Proc Natl Acad Sci U S A. 1998 Jul 21;95(15):8892–7.

Mortensen SA, et al. Coenzyme Q10: clinical benefits with biochemical correlates suggesting a scientific breakthrough in the management of chronic heart failure. Int J Tissue React. 1990;12(3):155–62.

Muroyama A. An alternative medical approach for the neuroprotective therapy to slow the progression of Parkinson's disease. Yakugaku Zasshi. 2013;133(8):849–56. doi:10.1248 /yakushi.13-00158.

Morris G, et al. Coenzyme Q10 depletion in medical and neuropsychiatric disorders: potential repercussions and therapeutic implications. Mol Neurobiol. 2013 Dec;48(3):883–903. Epub 2013 Jun 13. doi:10.1007/s12035-013-8477-8.

Nicolson GL. Mitochondrial dysfunction and chronic disease: treatment with natural supplements. Altern Ther Health Med. 2013 Aug 15. pii: at5027. Epub ahead of print.

Ochoa JJ, et al. Coenzyme Q10 protects from aging-related oxidative stress and improves mitochondrial function in heart of rats fed a polyunsaturated fatty acid (PUFA)-rich diet. J Gerontol A Biol Sci Med Sci. 2005 Aug;60(8):970–5. doi:10.1093/gerona/60.8.970.

Rodriguez MC, et al. Beneficial effects of creatine, CoQ10, and lipoic acid in mitochondrial disorders. Muscle Nerve. 2007 Feb;35(2):235–42. doi:10.1002/mus.20688.

Rosenfeldt FL, et al. Coenzyme Q10 improves the tolerance of the senescent myocardium to aerobic and ischemic stress: studies in rats and in human atrial tissue. Biofactors. 1999;9(2–4):291–9. doi:10.1002/biof.5520090226.

Rosenfeldt FL, et al. Coenzyme Q10 protects the aging heart against stress: studies in rats, human tissues, and patients. Ann N Y Acad Sci. 2002 Apr;959:355–9; discussion 463–5. doi:10.1111/j.1749-6632.2002.tb02106.x.

Rosenfeldt FL, et al. The effects of ageing on the response to cardiac surgery: protective strategies for the ageing myocardium. Biogerontology. 2002;3(1–2):37–40.

Salama M, et al. Co-enzyme Q10 to treat neurological disorders: basic mechanisms, clinical outcomes, and future research direction. CNS Neurol Disord Drug Targets. 2013 Aug;12(5):641–4. Epub 2013 Apr 4. doi:10.2174/18715273113129990071.

Shults CW, et al. Effects of coenzyme Q10 in early Parkinson disease: evidence of slowing of the functional decline. Arch Neurol. 2002 Oct;59(10):1541–50. doi:10.1001/archneur.59 .10.1541.

Sinatra ST. The Sinatra solution: metabolic cardiology. Laguna Beach, CA: Basic Health Publications, Inc; 2011.

Sohal RS, Forster MJ. Coenzyme Q, oxidative stress and aging. Mitochondrion. 2007 Jun;7 Suppl:S103–11. doi:10.1016/j.mito.2007.03.006.

Spindler M, Beal MF, Henchcliffe C. Coenzyme Q10 effects in neurodegenerative disease. Neuropsychiatr Dis Treat. 2009;5:597–610. Epub 2009 Nov 16. doi:10.2147/NDT.S5212.

CoQ10 as Adjunct to Statin Therapy

Brown, MS, inventor. Merck & Co., Inc., assignee. Coenzyme Q10 with HMG-CoA reductase inhibitors. United States patent US 4933165. 1989 Jan 18.

Caso G, et al. Effect of coenzyme Q10 on myopathic symptoms in patients treated with statins. Am J Cardiol. 2007 May 15;99(10):1409–12. doi:10.1016/j.amjcard.2006.12.063.

Marcoff L, Thompson PD. The role of coenzyme Q10 in statin-associated myopathy: a systematic review. J Am Coll Cardiol. 2007 Jun 12;49(23):2231–7. doi:10.1016/j.jacc.2007.02.049.

Parker BA, et al. Effect of statins on creatine kinase levels before and after a marathon run. Am J Cardiol. 2012 Jan 15;109(2):282–7. doi:10.1016/j.amjcard.2011.08.045.

Tobert JA, inventor. Merck & Co Inc., assignee. Coenzyme Q10 with HMG-CoA reductase inhibitors. United States patent US 4929437. 1990 May 29.

L-Carnitine

Akisu M, et al. Protective effect of dietary supplementation with L-arginine and L-carnitine on hypoxia/reoxygenation-induced necrotizing enterocolitis in young mice. Bio Neonate. 2002;81(4):260–5. doi:10.1159/000056757.

Bahl JJ, Bressler R. The pharmacology of carnitine. Annu Rev Pharmacol Toxicol. 1987;27: 257–77. doi:10.1146/annurev.pa.27.040187.001353.

Binienda ZK. Neuroprotective effects of L-carnitine in induced mitochondrial dysfunction. Ann N Y Acad Sci. 2003 May;993:289–95; discussion 345–9. doi:10.1111/j.1749-6632.2003.tb07536.

Brass EP, Hoppel CL. Relationship between acid-soluble carnitine and coenzyme A pools in vivo. Biochem J. 1980 Sep 15;190(3):495–504. doi:10.1042/bj1900495.

Bremer J. Carnitine: metabolism and functions. Physiol Rev. 1983 Oct;63(4):1420–80.

Ferrari R, et al. Therapeutic effects of L-carnitine and propionyl-L-carnitine on cardiovascular diseases: a review. Ann N Y Acad Sci. 2004 Nov;1033:79–91. doi:10.1196/annals.1320.007.

Geier DA, Geier MR. L-carnitine exposure and mitochondrial function in human neuronal cells. Neurochem Res. 2013 Nov;38(11):2336–41. Epub 2013 Sep 5. doi:10.1007/s11064-013 -1144-7.

Hagen TM, et al. Acetyl-L-carnitine fed to old rats partially restores mitochondrial function and ambulatory activity. Proc Natl Acad Sci U S A. 1998 Aug 4;95(16):9562–6.

Hagen TM, et al. Feeding acetyl-L-carnitine and lipoic acid to old rats significantly improves metabolic function while decreasing oxidative stress. Proc Natl Acad Sci U S A. 2002 Feb 19;99(4):1870–5. doi:10.1073/pnas.261708898.

Hoppel C. The role of carnitine in normal and altered fatty acid metabolism. Am J Kidney Dis. 2003 Apr; 41(4 Suppl 4):S4–12. doi:10.1016/S0272-6386(03)00112-4.

Horne DW, Broquist HP. Role of lysine and e-N-trimethyllysine in carnitine biosynthesis. I: studies in Neurospora crassa. J Biol Chem. 1973;248(6):2170–5.

Hulse JD, Ellis SR, Henderson LM. Carnitine biosynthesis: betahydroxylation of trimethyllysine by an alpha-ketoglutarate-dependent mitochondrial dioxygenase. J Biol Chem. 1978 Mar 10;253(5):1654–9.

Jacobs PL, Goldstein ER. Long-term glycine propionyl-l-carnitine supplemention and paradoxical effects on repeated anaerobic sprint performance. J Int Soc Sports Nutr. 2010 Oct 28;7:35. doi:10.1186/1550-2783-7-35.

Kabaroglu C, et al. Effects of L-arginine and L-carnitine on hypoxia/reoxygenation-induced intestinal injury. Pediatr Int. 2005 Feb;47(1):10–4. doi:10.1111/j.1442-200x.2005.01999.x.

Kuratsune H, et al. Acylcarnitine deficiency in chronic fatigue syndrome. Clin Infect Dis. 1994 Jan;18 Suppl 1:S62–7.

Lango R, et al. Propionyl-L-carnitine improves hemodynamics and metabolic markers of cardiac perfusion during coronary surgery in diabetic patients. Cardiovasc Drugs Ther. 2005 Aug;19(4):267–75.

Liu J, et al. Memory loss in old rats is associated with brain mitochondrial decay and RNA/ DNA oxidation: partial reversal by feeding acetyl-L-carnitine and/or R-α-lipoic acid. Proc Natl Acad Sci U S A. 2002 Feb 19;99(4):2356–61. doi:10.1073/pnas.261709299.

Lombard KA, et al. Carnitine status of lactoovovegetarians and strict vegetarian adults and children. Am J Clin Nutr. 1989;50(2):301–6.

McGarry JD, Brown NF. The mitochondrial carnitine palmitoyltransferase system. From concept to molecular analysis. Eur J Biochem. 1997 Feb 15;244(1):1–14. doi:10.1111/j .1432-1033.1997.00001.

Montgomery SA, Thal LJ, Amrein R. Meta-analysis of double blind randomized controlled clinical trials of acetyl-L-carnitine versus placebo in the treatment of mild cognitive impairment and mild Alzheimer's disease. Int Clin Psychopharmacol. 2003 Mar;18(2):61–71. doi:10.1097/01.yic.0000058280.28578.79.

Mortensen SA, et al. The effect of coenzyme Q10 on morbidity and mortality in chronic heart failure: results from the Q-SYMBIO: a randomized double-blind trial. JACC Heart Fail. 2014 Dec;2(6):641–9. doi:10.1016/j.jchf.2014.06.008.

Noland RC, et al. Carnitine insufficiency caused by aging overnutrition compromises mitochondrial performance and metabolic control. J Biol Chem. 2009 Aug 21;284(34): 22840–52. doi:10.1074/jbc.M109.032888.

Osmundsen H, Bremer J, Pedersen JI. Metabolic aspects of peroxisomal betaoxidation. Biochim Biophys Acta. 1991 Sep 11;1085(2):141–58. doi:10.1016/0005-2760(91)90089-Z.

Pande SV. A mitochondrial carnitine acylcarnitine translocase system. Proc Natl Acad Sci U S A. 1975 Mar;72(3):883–7.

Pande SV, Parvin R. Carnitine-acylcarnitine translocase catalyzes an equilibrating unidirectional transport as well. J Biol Chem. 1980 Apr 10;255(7):2994–3001.

Plioplys AV, Plioplys S. Serum levels of carnitine in chronic fatigue syndrome: clinical correlates. Neuropsychobiology 1995;32:132–8.

Pons R, De Vivo DC. Primary and secondary carnitine deficiency syndromes. J Child Neurol. 1995 Nov 1;10 Suppl 2:S8–24.

Ramsay RR, Arduini A. The carnitine acyltransferases and their role in modulating acyl-CoA pools. Arch Biochem Biophys. 1993 May;302(2):307–14. doi:10.1006/abbi.1993.1216.

Rebouche CJ. Kinetics, pharmacokinetics, and regulation of L-carnitine and acetyl-L-carnitine metabolism. Ann N Y Acad Sci. 2004 Nov;1033:30–41.

Rebouche CJ, Paulson DJ. Carnitine metabolism and function in humans. Annu Rev Nutr. 1986;6:41–66.

Rebouche CJ, Paulson DJ. Carnitine function and requirements during the life cycle. FASEB J. 1992; 6 (15):3379–86. doi:10.1146/annurev.nu.06.070186.000353.

Reuter SE, Evans AM. Carnitine and acylcarnitines: pharmacokinetic, pharmacological and clinical aspects. Clin Pharmacokinet. 2012 Sep 1;51(9):553–72. doi:10.2165/11633940 -000000000-00000.

Sachan DS, Broquist HP. Synthesis of carnitine from epsilon-N-trimethyllysine in post mitochondrial fractions of Neurospora crassa. Biochem Biophys Res Commun. 1980 Sep 30;96(2):870–5. doi:10.1016/0006-291X(80)91436-9.

Sachan DS, Hoppel CL. Carnitine biosynthesis. Hydroxylation of N-6-trimethyl-lysine to 3-hydroxy-N6-trimethyl-lysine. Biochem J. 1980 May 15; 188 (2):529–34. doi:10.1042/bj1880529.

Serati AR, et al. L-carnitine treatment in patients with mild diastolic heart failure is associated with improvement in diastolic function and symptoms. Cardiology. 2010;116(3):178–82. doi:10.1159/000318810.

Sinatra ST. The Sinatra solution: metabolic cardiology. Laguna Beach, CA: Basic Health Publications, Inc; 2011.

Steiber A, Kerner J, Hoppel CL. Carnitine: a nutritional, biosynthetic, and functional perspective. Mol Aspects Med. 2004 Oct–Dec;25(5–6):455–73. doi:10.1016/j.mam.2004.06.006.

Tanphaichitr V, Broquist HP. Role of lysine and e–N-trimethyllysine in carnitine biosynthesis. II: studies in the rat. J Biol Chem. 1973; 248(6):2176–81.

Vaz FM, Wanders RJ. Carnitine biosynthesis in mammals. Biochem J. 2002 Feb 1;361(Part 3): 417–29. doi:10.1042/bj3610417.

Virmani A, et al. The protective role of L-carnitine against neurotoxicity evoked by drug of abuse, methamphetamine, could be related to mitochondrial dysfunction. Ann N Y Acad Sci. 2002 Jun;965:225–32. doi:10.1111/j.1749–6632.2002.tb04164.

Magnesium

Abbott RD, et al. Dietary magnesium intake and the future risk of coronary heart disease (the Honolulu Heart Program). Am J Cardiol. 2003 Sep 15;92(6):665–9. doi:10.1016/S0002-9149(03)00819-1.

Alloui A, et al. Does Mg2+ deficiency induce a long-term sensitization of the central nociceptive pathways? Eur J Pharmacol. 2003 May 23;469(1–3)65–9. doi:10.1016/S0014-2999(03)01719-9.

Amighi J, et al. Low serum magnesium predicts neurological patients with advanced atherosclerosis. Stroke. 2004 Jan;35(1):22–7. doi:10.1161/01.STR.0000105928.95124.1F.

Demougeot C, et al. Effect of diets with different magnesium content in ischemic stroke rats. Neurosci Lett. 2004 May 13;362(1):17–20. doi:10.1016/j.neulet.2004.01.034.

Eray O, et al. Magnesium efficacy in magnesium deficient and nondeficient patients with rapid ventricular response atrial fibrillation. Eur J Emerg Med. 2000 Dec;7(4):287–90.

Fox C, Ramsoomair D, Carter C. Magnesium: its proven and potential clinical significance. South Med J. 2001 Dec;94(12):1195–201.

Hagen TM, et al. (R)-alpha-lipoic acid-supplemented old rats have improved mitochondrial function, decreased oxidative damage, and increased metabolic rate. FASEB J. 1999 Feb;13(2):411–8.

Hagen TM, et al. Mitochondrial decay in the aging rat heart: evidence for improvement by dietary supplementation with acetyl-L-carnitine and/or lipoic acid. Ann N Y Acad Sci. 2002 Apr;959:491–507. doi:10.1111/j.1749-6632.2002.tb02119.

Hans CP, Chaudhary DP, Bansal DD. Magnesium deficiency increases oxidative stress in rats. Ind J Exp Biol. 2002 Nov;40(11):1275–9.

Hans CP, Chaudhary DP, Bansal DD. Effect of magnesium supplementation on oxidative stress in alloxanic diabetic rats. Magnes Res. 2003 Mar;l6(1):13–9.

Klevay LM, Milne DB. Low dietary magnesium increases supraventricular ectopy. Am J Clin Nutr. 2002 Mar;75(3):550–4.

Kramer JH, et al. Dietary magnesium intake influences circulating proinflammatory neuropeptide levels and loss of myocardial tolerance to postischemic stress. Exp Biol Med (Maywood). 2003 Jun;228(6):665–73.

Kubota T, et al. Mitochondria are intracellular magnesium stores: investigation by simultaneous fluorescent imagings in PC12 cells. Biochim Biophys Acta. 2005 May 15;1744(1):19–28. Epub 2004 Nov 11. doi:10.1016/j.bbamcr.2004.10.013.

Laires MJ, Monteiro CP, Bicho M. Role of cellular magnesium in health and human disease. Front Biosci. 2004 Jan;9:262–76.

Lukaski HC, Nielsen FH. Dietary magnesium depletion affects metabolic responses during submaximal exercise in postmenopausal women. J Nutr. 2002 May;132(5):930–5.

Maier JA, et al. Low magnesium promotes endothelial cell dysfunction: implications for atherosclerosis, inflammation and thrombosis. Biochim Biophys Acta. 2004 May 24;1689(l):13–21. doi:10.1016/j.bbadis.2004.01.002.

Moreira PI, et al. Lipoic acid and N-acetyl cysteine decrease mitochondrial-related oxidative stress in Alzheimer disease patient fibroblasts. J Alzheimers Dis. 2007 Sep;12(2):195–206.

Nair RR, Nair P. Alteration of myocardial mechanics in marginal magnesium deficiency. Magnes Res. 2002 Dec;15(3–4):287–306.

Nakayama S, et al. Mechanisms for monovalent cation-dependent depletion of intracellular Mg2+:Na(+)-independent Mg2+ pathways in guinea-pig smooth muscle. J Physiol. 2003 Sep 15;551(Pt 3):843–53. doi:10.1113/jphysiol.2003.047795.

Paolisso G, Barbagallo M. Hypertension, diabetes mellitus, and insulin resistance: the role of intracellular magnesium. Am J Hyperten. 1997 Mar 1;10(3):346–55. doi:10.1016/S0895 -7061(96)00342-1.

Resnick LM, et al. Cellular-free magnesium depletion in brain and muscle of normal and preeclamptic pregnancy: a nuclear magnetic resonance spectroscopic study. Hypertension. 2004 Sep;44(3):322–6. doi:10.1161/01.HYP.0000137592.76535.8c.

Rubenowitz E, Axelsson G, Rylander R. Magnesium in drinking water and death from myocardial infarction. Am J Epidemiol. l996;143:456–62.

Sinatra ST. The Sinatra solution: metabolic cardiology. Laguna Beach, CA: Basic Health Publications, Inc; 2011.

Takaya J, Higashino H, Kobayashi Y. Intracellular magnesium and insulin resistance. Magnes Res. 2004 Jun;17(2):126–36.

Touyz RM. Role of magnesium in the pathogenesis of hypertension. Mol Aspects Med. 2003 Feb–Jun;24(1–3):107–36. doi:10.1016/S0098-2997(02)00094-8.

Touyz RM, et al. Effects of low dietary magnesium intake on development of hypertension in stroke-prone spontaneously hypertensive rats: role of reactive oxygen species. J Hypertens. 2002 Nov;20(11):2221–32.

Alpha-Lipoic Acid

Biewenga GP, Haenen GR, Bast A. The pharmacology of the antioxidant lipoic acid. Gen Pharmacol. 1997 Sep;29(3):315–31. doi:10.1016/S0306-3623(96)00474-0.

Femiano F, Scully C. Burning mouth syndrome (BMS): double blind controlled study of alphalipoic acid (thioctic acid) therapy. J Oral Pathol Med. 2002 May;31(5):267–9. doi:10 .1034/j.1600-0714.2002.310503.

Hagen TM, et al. (R)-alpha-lipoic acid-supplemented old rats have improved mitochondrial function, decreased oxidative damage, and increased metabolic rate. FASEB J. 1999 Feb;13(2):411–8.

Hagen TM, et al. Feeding acetyl-L-carnitine and lipoic acid to old rats significantly improves metabolic function while decreasing oxidative stress. Proc Natl Acad Sci U S A. 2002 Feb 19;99(4):1870–5. doi:10.1073/pnas.261708898.

Hagen TM, et al. Mitochondrial decay in the aging rat heart: evidence for improvement by dietary supplementation with acetyl-L-carnitine and/or lipoic acid. Ann N Y Acad Sci. 2002 Apr;959:491–507. doi:10.1111/j.1749-6632.2002.tb02119.

Hager K, et al. Alpha-lipoic acid as a new treatment option for Alzheimer type dementia. Arch Gerontol Geriatr. 2001 Jun;32(3):275–82.

Jia L, et al. Acrolein, a toxicant in cigarette smoke, causes oxidative damage and mitochondrial dysfunction in RPE cells: protection by (R)-alpha-lipoic acid. Invest Ophthalmol Vis Sci. 2007 Jan; 48(1):339–48. doi:10.1167/iovs.06-0248.

Jiang T, et al. Lipoic acid restores age-associated impairment of brain energy metabolism through the modulation of Akt/JNK signaling and PGC1α transcriptional pathway. Aging Cell. 2013 Dec;12(6):1021–31. doi:10.1111/acel.12127. doi:10.1111/acel.12127.

Kim DC, et al. Lipoic acid prevents the changes of intracellular lipid partitioning by free fatty acid. Gut Liver. 2013 Mar;7(2):221–7. doi:10.5009/gnl.2013.7.2.221.

Li CJ, et al. Attenuation of myocardial apoptosis by alpha-lipoic acid through suppression of mitochondrial oxidative stress to reduce diabetic cardiomyopathy. Chin Med J (Engl). 2009 Nov 5;122(21):2580–6.

Liu J, Killilea DW, Ames BN. Age-associated mitochondrial oxidative decay: improvement of carnitine acetyltransferase substrate-binding affinity and activity in brain by feeding old rats acetyl-L-carnitine and/or R-alpha-lipoic acid. Proc Natl Acad Sci U S A. 2002 Feb 19;99(4):1876–81. doi:10.1073/pnas.261709098.

Liu J, et al. Delaying brain mitochondrial decay and aging with mitochondrial antioxidants and metabolites. Ann N Y Acad Sci. 2002 Apr;959:133–66. doi:10.1111/j.1749-6632.2002.tb02090.x.

Liu J, et al. Memory loss in old rats is associated with brain mitochondrial decay and RNA/ DNA oxidation: partial reversal by feeding acetyl-L-carnitine and/or R-alpha- lipoic acid. Proc Natl Acad Sci U S A. 2002 Feb 19;99(4):2356–61. doi:10.1073/pnas.261709299.

Meydani M, et al. The effect of long-term dietary supplementation with antioxidants. Ann N Y Acad Sci. 1998 Nov 20;854:352–60. doi:10.1111/j.1749-6632.1998.tb09915.

Nyengaard JR, et al. Interactions between hyperglycemia and hypoxia: implications for diabetic retinopathy. Diabetes. 2004 Nov;53(11):2931–8. doi:10.2337/diabetes.53.11.2931.

Scott BC, et al. Lipoic and dihydrolipoic acids as antioxidants. A critical evaluation. Free Radic Res. 1994 Feb;20(2):119–33. doi:10.3109/10715769409147509.

Tappel A, Fletcher B, Deamer D. Effect of antioxidants and nutrients on lipid peroxidation fluorescent products and aging parameters in the mouse. J Gerontol. 1973 Oct;28(4):415–24. doi:10.1093/geronj/28.4.415.

Thornalley PJ. Glycation in diabetic neuropathy: characteristics, consequences, causes, and therapeutic options. Int Rev Neurobiol. 2002;50:37–7. doi:10.1016/S0074-7742(02)50072-6.

Williamson JR, et al. Hyperglycemic pseudohypoxia and diabetic complications. Diabetes. 1993 Jun;42(6):801–13. doi:10.2337/diab.42.6.801.

Ziegler D, et al. Treatment of symptomatic diabetic polyneuropathy with the antioxidant alpha-lipoic acid: a meta-analysis. Diabet Med. 2004 Feb;21(2):114–21. doi:10.1111/j.1464-5491 .2004.01109.

Zhou L, et al. α-Lipoic acid ameliorates mitochondrial impairment and reverses apoptosis in FABP3-overexpressing embryonic cancer cells. J Bioenerg Biomembr. 2013 Oct;45(5): 459–66. Epub 2013 Mar 28.

Creatine

Andrews R, et al. The effect of dietary creatine supplementation on skeletal muscle metabolism in congestive heart failure. Eur Heart J. 1998 Apr;19(4):617–22.

Balestrino M, et al. Role of creatine and phosphocreatine in neuronal protection from anoxic and ischemic damage. Amino Acids. 2002; 23(1–3):221–9. doi:10.1007/s00726-001-0133-3.

Broqvist M, et al. Nutritional assessment and muscle energy metabolism in severe chronic congestive heart failure—effects of long-term dietary supplementation. Eur Heart J. 1994 Dec;15(12):1641–50. doi:10.1093/oxfordjournals.eurheartj.a060447.

Ferrante RJ, et al. Neuroprotective effects of creatine in a transgenic mouse model of Huntington's disease. J Neurosci. 2000 Jun 15;20(12):4389–97.

Field ML. Creatine supplementation in congestive heart failure [letter]. Cardiovasc Res 1996 Jan; 31(1):174–6.

Gordon A, et al. Creatine supplementation in chronic heart failure increases skeletal muscle creatine phosphate and muscle performance. Cardiovasc Res. 1995 Sep;30(3):413–8.

Klivenyi P, et al. Neuroprotective effects of creatine in a transgenic animal model of amyotrophic lateral sclerosis. Nat Med. 1999 Mar;5(3):347–50. doi:10.1038/6568.

Malcon C, Kaddurah-Daouk R, Beal MF. Neuroprotective effects of creatine administration against NMDA and malonate toxicity. Brain Res. 2000 Mar 31;860(1–2):195–8. doi:10.1016 /S0006-8993(00)02038-2.

Matthews RT, et al. Neuroprotective effects of creatine and cyclocreatine in animal models of Huntington's disease. J Neurosci. 1998 Jan;18(1):156–63.

Matthews RT, et al. Creatine and cyclocreatine attenuate MPTP neurotoxicity. Exp Neurol. 1999 May;157(1):142–9. doi:10.1006/exnr.1999.7049.

Park JH, et al. Use of P-31 magnetic resonance spectroscopy to detect metabolic abnormalities in muscles of patients with fibromyalgia. Arthritis Rheum. 1998 Mar;41(3):406–13. doi:10.1002/1529-0131(199803)41:3<406::AID-ART5>3.0.CO;2-L.

Tarnopolsky M, Martin J. Creatine monohydrate increases strength in patients with neuromuscular disease. Neurology. 1999 Mar 10;52(4):854–7.

Walter MC, et al. Creatine monohydrate in muscular dystrophies: a double blind, placebo-controlled clinical study. Neurology. 2000 May 9; 54(9):1848–50.

B Vitamins

Bernsen PL, et al. Successful treatment of pure myopathy, associated with complex I deficiency, with riboflavin and carnitine. Arch Neurol. 1991 Mar;48(3):334–8. doi:10.1001/archneur.1991.00530150106028.

Bernsen PL, et al. Treatment of complex I deficiency with riboflavin. J Neurol Sci. 1993 Sep;118(2):181–7. doi:10.1016/0022-510X(93)90108-B.

Bettendorff L, et al. Low thiamine diphosphate levels in brains of patients with frontal lobe degeneration of the non-Alzheimer's type. J Neurochem. 1997 Nov;69(5):2005–10. doi:10.1046/j.1471-4159.1997.69052005.

Bogan KL, Brenner C. Nicotinic acid, nicotinamide, and nicotinamide riboside: a molecular evaluation of NAD+ precursor vitamins in human nutrition. Annu Rev Nutr. 2008;28:115–30. doi:10.1146/annurev.nutr.28.061807.155443.

Bottiglieri T. Folate, vitamin B12, and s-adenosylmethionine. Psychiatr Clin North Am. 2013 Mar;36(1):1–13. doi:10.1016/j.psc.2012.12.001.

Bugiani M, et al. Effects of riboflavin in children with complex II deficiency. Brain Dev. 2006. Oct;28(9):576–81. doi:10.1016/j.braindev.2006.04.001.

Cantó C, et al. The NAD(+) precursor nicotinamide riboside enhances oxidative metabolism and protects against high-fat diet-induced obesity. Cell Metab. 2012 Jun 6;15(6):838–47. doi:10.1016/j.cmet.2012.04.022.

Czerniecki J, Czygier M. Cooperation of divalent ions and thiamin diphosphate in regulation of the function of pig heart pyruvate dehydrogenase complex. J Nutr Sci Vitaminol (Tokyo). 2001 Dec;47(6):385–6.

Denu JM. Vitamins and aging: pathways to NAD+ synthesis. Cell. 2007 May 4;129(3):453–4. doi:10.1016/j.cell.2007.04.023.

Garbin L, Plebani M, Terribile PM. Effect of ACP (pyridoxine-2-oxoglutarate) on CCl4 intoxication and in streptozotocin-induced ketosis in rat. Acta Vitaminol Enzymol. 1977;31(6):175–8.

Gerards M, et al. Riboflavin-responsive oxidative phosphorylation complex I deficiency caused by defective ACAD9: new function for an old gene. Brain. 2011 Jan;134(Pt 1):210–9. doi:10.1093/brain/awq273.

Hartman TJ, et al. Association of the B-vitamins pyridoxal 5'-phosphate (B(6)), B(12), and folate with lung cancer risk in older men. Am J Epidemiol. 2001 Apr 1;153(7):688–94. doi:10.1093/aje/153.7.688.

Iliev IS, et al. Enzyme activity changes in chronic alcoholic intoxication and the simultaneous administration of pyridoxine. Vopr Pitan. 1982 Nov;(6):54–6.

Imai SI, Guarente L. NAD(+) and sirtuins in aging and disease. Trends Cell Biol. 2014 Aug;24(8):464–71. Epub 2014 Apr 28. doi:10.1016/j.tcb.2014.04.002.

Ke ZJ, et al. Reversal of thiamine deficiency-induced neurodegeneration. J Neuropathol Exp Neurol. 2003 Feb;62(2):195–207. doi:10.1093/jnen/62.2.195.

Kelly G. The coenzyme forms of vitamin B12: toward an understanding of their therapeutic potential. Altern Med Rev. 1997 Sep;2(6):459–71.

Kotegawa M, Sugiyama M, Haramaki N. Protective effects of riboflavin and its derivatives against ischemic reperfused damage of rat heart. Biochem Mol Biol Int. 1994 Oct;34(4):685–91.

Maassen JA. Mitochondrial diabetes, diabetes and the thiamine-responsive megaloblastic anaemia syndrome and MODY-2. Diseases with common pathophysiology? Panminerva Med. 2002 Dec;44(4):295–300.

Magni G, et al. Enzymology of mammalian NAD metabolism in health and disease. Front Biosci. 2008 May 1;13:6135–54.

Marriage B, Clandinin MT, Glerum DM. Nutritional cofactor treatment in mitochondrial disorders. J Am Diet Assoc. 2003 Aug;103(8):1029–38. doi:10.1053/jada.2003.50196.

McComsey GA, Lederman MM. High doses of riboflavin and thiamine may help in secondary prevention of hyperlactatemia. AIDS Read. 2002 May;12(5):222–4.

Miner SE, et al. Pyridoxine improves endothelial function in cardiac transplant recipients. J Heart Lung Transplant. 2001 Sep;20(9):964–9. doi:10.1016/S1053-2498(01)00293-5.

Naito E, et al. Thiamine-responsive pyruvate dehydrogenase deficiency in two patients caused by a point mutation (F205L and L216F) within the thiamine pyrophosphate binding region. Biochim Biophys Acta. 2002 Oct 9;1588(1):79–84. doi:10.1016/S0925-4439(02)00142-4.

Okada H, et al. Vitamin B6 supplementation can improve peripheral polyneuropathy in patients with chronic renal failure on high-flux haemodialysis and human recombinant erythropoietin. Nephrol Dial Transplant. 2000 Sep;15(9):1410–3. doi:10.1093/ndt/15.9.1410.

Pomero F, et al. Benfotiamine is similar to thiamine in correcting endothelial cell defects induced by high glucose. Acta Diabetol. 2001;38(3):135–8.

Sasaki Y, Araki T, Milbrandt J. Stimulation of nicotinamide adenine dinucleotide biosynthetic pathways delays axonal degeneration after axotomy. J Neurosci. 2006 Aug 16;26(33):8484–91. doi:10.1523/JNEUROSCI.2320-06.2006.

Sato Y, et al. Mitochondrial myopathy and familial thiamine deficiency. Muscle Nerve. 2000 Jul;23(7):1069–75. doi:10.1002/1097-4598(200007)23:7<1069::AID-MUS9>3.0.CO;2-0.

Sauve AA. NAD+ and vitamin B3: from metabolism to therapies. J Pharmacol Exp Ther. 2008 Mar;324(3):883–93. Epub 2007 Dec 28. doi:10.1124/jpet.107.120758.

Scholte HR, et al. Riboflavin-responsive complex I deficiency. Biochim Biophys Acta. 1995 May 24;1271(1):75–83. doi:10.1016/0925-4439(95)00013-T.

Sheline CT, et al. Cofactors of mitochondrial enzymes attenuate copper-induced death in vitro and in vivo. Ann Neurol. 2002 Aug;52(2):195–204. doi:10.1002/ana.10276.

Subramanian VS, et al. Mitochondrial uptake of thiamin pyrophosphate: physiological and cell biological aspects. PLoS One. 2013 Aug 30;8(8):e73503. doi:10.1371/journal.pone.0073503.

Tahiliani AG, Beinlich CJ. Pantothenic acid in health and disease. Vitam Horm. 1991 Feb;46:165–228. doi:10.1016/S0083-6729(08)60684-6.

Tempel W, et al. Nicotinamide riboside kinase structures reveal new pathways to NAD+. PLoS Biol. 2007 Oct 2;5(10):e263. doi:10.1371/journal.pbio.0050263

Togay-Isikay C, Yigit A, Mutluer N. Wernicke's encephalopathy due to hyperemesis gravidarum: an under-recognised condition. Aust N Z J Obstet Gynaecol. 2001 Nov;41(4):453–6. doi:10.1111/j.1479-828X.2001.tb01330.

Watanabe F. Vitamin B12 sources and bioavailability. Exp Biol Med (Maywood). 2007 Nov;232(10):1266–74. doi:10.3181/0703-MR-67.

Yang T, Chan NY, Sauve AA. Syntheses of nicotinamide riboside and derivatives: effective agents for increasing nicotinamide adenine dinucleotide concentrations in mammalian cells. J Med Chem. 2007 Dec 27;50(26):6458–61. Epub 2007 Dec 6. doi:10.1021/jm701001c.

Youssef JA, Song WO, Badr MZ. Mitochondrial, but not peroxisomal, beta-oxidation of fatty acids is conserved in coenzyme A-deficient rat liver. Mol Cell Biochem. 1997 Oct;175(1–2):37–42.

Iron

Atamna H. Heme, iron, and the mitochondrial decay of ageing. Aging Res Rev. 2004 Jul;3(3):303–18. doi:10.1016/j.arr.2004.02.002

Atamna H, et al. Heme deficiency may be a factor in the mitochondrial and neuronal decay of aging. Proc Natl Acad Sci U S A. 2002 Nov 12;99(23):14807–12. Epub 2002 Nov 4. doi:10.1073/pnas.192585799.

Stoltzfus RJ. Iron deficiency: global prevalence and consequences. Food Nutr Bull. 2003 Dec;24(4 Suppl):S99–103. doi:10.1177/15648265030244S206.

Resveratrol and Pterostilbene

Alcaín FJ, Villalba JM. Sirtuin activators. Expert Opin Ther Pat. 2009 Apr;19(4):403–14. doi:10.1517/13543770902762893.

Alosi JA, et al. Pterostilbene inhibits breast cancer in vitro through mitochondrial depolarization and induction of caspase-dependent apoptosis. J Surg Res. 2010 Jun 15;161(2):195–201. Epub 2009 Aug 18. doi:10.1016/j.jss.2009.07.027.

Bagchi D, et al. Molecular mechanisms of cardioprotection by a novel grape seed proanthocyanidin extract. Mutat Res. 2003 Feb–Mar;523–524:87–97. doi:10.1016/S0027-5107(02)00324-X.

Baur JA, et al. Resveratrol improves health and survival of mice on a high-calorie diet. Nature. 2006 Nov 16;444(7177):337–42. doi:10.1038/nature05354.

Chiou YS, et al. Pterostilbene is more potent than resveratrol in preventing azoxymethane (AOM)-induced colon tumorigenesis via activation of the NF-E2-related factor 2 (Nrf2)-mediated antioxidant signaling pathway. J Agric Food Chem. 2011 Mar 23;59(6):2725–33. Epub 2011 Feb 28. doi:10.1021/jf2000103.

Joseph JA, et al. Cellular and behavioral effects of stilbene resveratrol analogues: implications for reducing the deleterious effects of aging. J Agric Food Chem. 2008;56(22):10544–51. doi:10.1021/jf802279h.

Lagouge M, et al. Resveratrol improves mitochondrial function and protects against metabolic disease by activating SIRT1 and PGC-1alpha. Cell. 2006 Dec 15;127(6):1109–22. doi:10.1016/j.cell.2006.11.013.

Li YG, et al. Resveratrol protects cardiomyocytes from oxidative stress through SIRT1 and mitochondrial biogenesis signaling pathways. Biochem Biophys Res Commun. 2013 Aug 23;438(2):270–6. Epub 2013 Jul 24. doi:10.1016/j.bbrc.2013.07.042.

Lin VC, et al. Activation of AMPK by pterostilbene suppresses lipogenesis and cell-cycle progression in p53 positive and negative human prostate cancer cells. J Agric Food Chem. 2012 Jun 27;60(25):6399–407. Epub 2012 Jun 19. doi:10.1021/jf301499e.

Macicková T, et al. Effect of stilbene derivative on superoxide generation and enzyme release from human neutrophils in vitro. Interdiscip Toxicol. 2012 Jun;5(2):71–5. doi:10.2478/v10102-012-0012-7.

Meng XL, et al. Effects of resveratrol and its derivatives on lipopolysaccharide-induced microglial activation and their structure-activity relationships. Chem Biol Interact. 2008 Jul 10;174(1):51–9. doi:10.1016/j.cbi.2008.04.015.

Moon D, et al. Pterostilbene induces mitochondrially derived apoptosis in breast cancer cells in vitro. J Surg Res. 2013 Apr;180(2):208–15. Epub 2012 Apr 29. doi:10.1016/j.jss.2012.04.027.

Nutakul W, et al. Inhibitory effects of resveratrol and pterostilbene on human colon cancer cells: a side-by-side comparison. J Agric Food Chem. 2011 Oct 26;59(20):10964–70. doi:10.1021/jf202846b.

Pan MH, et al. Pterostilbene induces apoptosis and cell cycle arrest in human gastric carcinoma cells. J Agric Food Chem. 2007 Sep 19;55(19):7777–85. Epub 2007 Aug 16. doi:10.1021/jf071520h.

Pan Z, et al. Identification of molecular pathways affected by pterostilbene, a natural dimethylether analog of resveratrol. BMC Med Genomics. 2008 Mar 20;1:7. doi:10.1186/1755-8794-1-7.

Pari L, Satheesh MA. Effect of pterostilbene on hepatic key enzymes of glucose metabolism in streptozotocin- and nicotinamide-induced diabetic rats. Life Sci. 2006;79(7):641–5. doi:10.1016/j.lfs.2006.02.036.

Park ES, et al. Pterostilbene, a natural dimethylated analog of resveratrol, inhibits rat aortic vascular smooth muscle cell proliferation by blocking Akt-dependent pathway. Vascul Pharmacol. 2010 Jul–Aug;53(1–2):61–7. doi:10.1016/j.vph.2010.04.001.

Pearson KJ, et al. Resveratrol delays age-related deterioration and mimics transcriptional aspects of dietary restriction without extending lifespan. Cell Metab. 2008 Aug;8(2):157–68. doi:10.1016/j.cmet.2008.06.011.

Polley KR, et al. Influence of exercise training with resveratrol supplementation on skeletal muscle mitochondrial capacity. Appl Physiol Nutr Metab. 2016;41(1):26–32. doi:10.1139/apnm-2015-0370.

Priego S, et al. Natural polyphenols facilitate elimination of HT-29 colorectal cancer xenografts by chemoradiotherapy: a Bcl-2- and superoxide dismutase 2-dependent mechanism. Mol Cancer Ther. 2008 Oct;7(10):3330–42. doi:10.1158/1535-7163.MCT-08-0363.

Remsberg CM, et al. Pharmacometrics of pterostilbene: preclinical pharmacokinetics and metabolism, anticancer, antiinflammatory, antioxidant and analgesic activity. Phytother Res. 2008 Feb;22(2):169–79. doi:10.1002/ptr.2277.

Rimando AM, et al. Pterostilbene, a new agonist for the peroxisome proliferator-activated receptor r-Isoform, lowers plasma lipoproteins and cholesterol in hypercholesterolemic hamsters. J Agric Food Chem. 2005;53:3403–7. doi:10.1021/jf0580364.

Siva B, et al. Effect of polyphenoics extracts of grape seeds (GSE) on blood pressure (BP) in patients with the metabolic syndrome (MetS). FASEB J. 2006;20:A305.

Wang J, et al. Grape-derived polyphenolics prevent Abeta oligomerization and attenuate cognitive deterioration in a mouse model of Alzheimer's disease. J Neurosci. 2008 Jun 18;28(25):6388–92. doi:10.1523/JNEUROSCI.0364-08.2008.

Williams CM, et al. Blueberry-induced changes in spatial working memory correlate with changes in hippocampal CREB phosphorylation and brain-derived neurotrophic factor (BDNF) levels. Free Radic Biol Med. 2008 Aug 1;45(3):295–305. doi:10.1016/j.freeradbiomed.2008.04.008.

Youdim KA, et al. Short-term dietary supplementation of blueberry polyphenolics: beneficial effects on aging brain performance and peripheral tissue function. Nutr Neurosci. 2000 Jul 13;3:383–97. doi:10.1080/1028415X.2000.11747338.

Ketogenic Diets and Calorie Restriction

Anderson RM, et al. Manipulation of a nuclear NAD+ salvage pathway delays aging without altering steady-state NAD+ levels. J Biol Chem. 2002 May 24;277(21):18881–90. doi:10.1074/jbc.M111773200.

Araki T, Sasaki Y, Milbrandt J. Increased nuclear NAD biosynthesis and SIRT1 activation prevent axonal degeneration. Science. 2004 Aug 13;305(5686):1010–3. doi:10.1126/science.1098014.

Campbell MK, Farrell, SO. Biochemistry. 5th edition. Pacific Grove, Thomson Brooks/Cole; 2006. 579 p.

Carrière A, et al. Browning of white adipose cells by intermediate metabolites: an adaptive mechanism to alleviate redox pressure. Diabetes. 2014 Oct;63(10):3253–65. Epub 2014 May 1. doi:10.2337/db13-1885.

Castello L, et al. Calorie restriction protects against age-related rat aorta sclerosis. FASEB J. 2005 Nov;19(13):1863–5. Epub 2005 Sep 8. doi:10.1096/fj.04-2864fje.

Cohen HY, et al. Calorie restriction promotes mammalian cell survival by inducing the SIRT1 deacetylase. Science. 2004 Jul 16;305(5682):390–2. doi:10.1126/science.1099196.

Colman RJ, et al. Caloric restriction reduces age-related and all-cause mortality in rhesus monkeys. Nat Commun. 2014 Apr 1;5:3557. doi:10.1038/ncomms4557.

Estrada NM, Isokawa M. Metabolic demand stimulates CREB signaling in the limbic cortex: implication for the induction of hippocampal synaptic plasticity by intrinsic stimulus for survival. Front Syst Neurosci. 2009 Jun 9;3:5. doi:10.3389/neuro.06.005.2009.

Guarente L, Picard F. Calorie restriction — the SIR2 connection. Cell. 2005 Feb 25;120(4): 473–82. doi:10.1016/j.cell.2005.01.029.

Hasselbalch SG, et al. Brain metabolism during short-term starvation in humans. J Cereb Blood Flow Metab. 1994 Jan;14(1):125–31. doi:10.1038/jcbfm.1994.17.

Ivanova DG, Yankova TM. The free radical theory of aging in search of a strategy for increasing life span. Folia Med (Plovdiv). 2013 Jan–Mar;55(1):33–41. doi:10.2478/folmed-2013-0003.

Jarrett SG, et al. The ketogenic diet increases mitochondrial glutathione levels. J Neurochem. 2008 Aug;106(3):1044–51. doi:10.1111/j.1471-4159.2008.05460.x. Epub 2008 May 5.

Jung KJ, et al. The redox-sensitive DNA binding sites responsible for age-related downregulation of SMP30 by ERK pathway and reversal by calorie restriction. Antioxid Redox Signal. 2006 Mar–Apr;8(3–4):671–80. doi:10.1089/ars.2006.8.671.

Kashiwaya Y, et al. D-b-hydroxybutyrate protects neurons in models of Alzheimer's and Parkinson's disease. Proc Natl Acad Sci U S A. 2000 May 9;97(10):5440–4. doi:10.1073/pnas.97.10.5440.

Kodde IF, et al. Metabolic and genetic regulation of cardiac energy substrate preference. Comp Biochem Physiol A Mol Integr Physiol. 2007 Jan;146(1):26–39. Epub 2006 Oct 3. doi:10.1016/j.cbpa.2006.09.014.

Laffel L. Ketone bodies: a review of physiology, pathophysiology and application of monitoring to diabetes. Diabetes Metab Res Rev. 1999 Nov–Dec;15(6):412–26. doi:10.1002 /(SICI)1520-7560(199911/12)15:6<412::AID-DMRR72>3.0.CO;2-8.

Lim EL, et al. Reversal of type 2 diabetes: normalisation of beta cell function in association with decreased pancreas and liver triacylglycerol. Diabetologia. 2011 Oct;54(10):2506–14. Epub 2011 Jun 9. doi:10.1007/s00125-011-2204-7.

Lin SJ, et al. Calorie restriction extends yeast life span by lowering the level of NADH. Genes Dev. 2004 Jan 1;18(1):12–6. doi:10.1101/gad.1164804.

Mattson MP, Chan SL, Duan W. Modification of brain aging and neurodegenerative disorders by genes, diet, and behavior. Physiol Rev. 2002 Jul;82(3):637–72. doi:10.1152/physrev.00004.2002.

McInnes N, et al. Piloting a remission strategy in type 2 diabetes: results of a randomized controlled trial. J Clin Endocrinol Metab. 2017 May 1;102(5):1596–1605. Epub 2017 Mar 15. doi:10.1210/jc.2016-3373.

Mercken EM, et al. Calorie restriction in humans inhibits the PI3K/AKT pathway and induces a younger transcription profile. Aging Cell. 2013 Aug;12(4):645–51. Epub 2013 Apr 20. doi:10.1111/acel.12088.

Picard F, et al. Sirt1 promotes fat mobilization in white adipocytes by repressing PPARgamma. Nature. 2004 Jun 17;429(6993):771–6. doi:10.1038/nature02583.

Prins ML. Cerebral metabolic adaptation and ketone metabolism after brain injury. J Cereb Blood Flow Metab. 2008 Jan;28(1):1–16. Epub 2007 Aug 8. doi:10.1038/sj.jcbfm .9600543.

Revollo JR, Grimm AA, Imai S. The NAD biosynthesis pathway mediated by nicotinamide phosphoribosyltransferase regulates Sir2 activity in mammalian cells. J Biol Chem. 2004 Dec 3;279(49):50754–63. doi:10.1074/jbc.M408388200.

Rose G, et al. Variability of the SIRT3 gene, human silent information regulator Sir2 homologue, and survivorship in the elderly. Exp Gerontol. 2003 Oct;38(10):1065–70. doi:10.1016/S0531-5565(03)00209-2.

Sato K, et al. Insulin, ketone bodies, and mitochondrial energy transduction. FASEB J. 1995 May;9(8):651–8.

Scheibye-Knudsen M, et al. A high-fat diet and NAD(+) activate Sirt1 to rescue premature aging in cockayne syndrome. Cell Metab. 2014 Nov 4;20(5):840–55. Epub 2014 Nov 4. doi:10.1016/j.cmet.2014.10.005.

Sharman MJ, et al. A ketogenic diet favorably affects serum biomarkers for cardiovascular disease in normal-weight young men. J Nutr. 2002 Jul;132(7):1879–85.

Sort R, et al. Ketogenic diet in 3 cases of childhood refractory status epilepticus. Eur J Paediatr Neurol. 2013 Nov;17(6):531–6. Epub 2013 Jun 7. doi:10.1016/j.ejpn.2013.05.001.

Spindler SR. Caloric restriction: from soup to nuts. Aging Res Rev. 2010 Jul;9(3):324–53. doi:10.1016/j.arr.2009.10.003.

VanItallie TB, Nufert TH. Ketones: metabolism's ugly duckling. Nutr Rev. 2003 Oct;61(10): 327–41. doi:10.1301/nr.2003.oct.327-341.

Veech RL, et al. Ketone bodies, potential therapeutic uses. IUBMB Life. 2001 Apr;51(4): 241–7. doi:10.1080/152165401753311780.

Wang SP, et al. Metabolism as a tool for understanding human brain evolution: lipid energy metabolism as an example. J Hum Evol. 2014 Dec;77:41–9. Epub 2014 Dec 6. doi:10.1016 /j.jhevol.2014.06.013.

Wegman MP, et al. Practicality of intermittent fasting in humans and its effect on oxidative stress and genes related to aging and metabolism. Rejuvenation Res. 2015 Apr;18(2):162–72. doi:10.1089/rej.2014.1624.

Wood JG, et al. Sirtuin activators mimic caloric restriction and delay aging in metazoans. Nature. 2004 Aug 5;430(7000):686–9. doi:10.1038/nature02789.

Massage and Hydrotherapy

Boon MR, et al. Brown adipose tissue: the body's own weapon against obesity? Ned Tijdschr Geneeskd. 2013;157(20):A5502.

Crane JD, et al. Massage therapy attenuates inflammatory signaling after exercise-induced muscle damage. Sci Transl Med. 2012 Feb 1;4(119):119ra13. doi:10.1126/scitranslmed .3002882.

Lee P, et al. Temperature-acclimated brown adipose tissue modulates insulin sensitivity in humans. Diabetes. 2014 Nov;63(11):3686–98. Epub 2014 Jun 22. doi:10.2337/db14-0513.

Lo KA, Sun L. Turning WAT into BAT: a review on regulators controlling the browning of white adipocytes. Biosci Rep. 2013 Sep 6;33(5):e00065. Epub Jul 30.

van der Lans AA, et al. Cold acclimation recruits human brown fat and increases nonshivering thermogenesis. J Clin Invest. 2013 Aug;123(8):3395–403. Epub 2013 Jul 15. doi:10.1172 /JCI68993.

Cannabis and Phytocannabinoids

Athanasiou A, et al. Cannabinoid receptor agonists are mitochondrial inhibitors: a unified hypothesis of how cannabinoids modulate mitochondrial function and induce cell death. Biochem Biophys Res Commun. 2007 Dec 7;364(1):131–7. doi:10.1016/j.bbrc.2007.09.107.

Bénard G, et al. Mitochondrial CB1 receptors regulate neuronal energy metabolism. Nat Neurosci. 2012 Mar 4;15(4):558–64. doi:10.1038/nn.3053.

Biophysical Society. Imbalance of calcium in a cell's energy factory may drive Alzheimer's disease. ScienceDaily. 2017 Feb 14. Available from: www.sciencedaily.com/releases/2017/02/170214172757.htm.

Cao C, et al. The potential therapeutic effects of THC on Alzheimer's disease. J Alzheimers Dis. 2014;42(3):973–84. doi:10.3233/JAD-140093.

Hao E, et al. Cannabidiol protects against doxorubicin-induced cardiomyopathy by modulating mitochondrial function and biogenesis. Mol Med. 2015 Jan 6;21:38–45. doi:10.2119/molmed.2014.00261.

Hebert-Chatelain E, et al. Cannabinoid control of brain bioenergetics: exploring the subcellular localization of the CB1 receptor. Mol Metab. 2014 Apr 2;3(4):495–504. doi:10.1016/j.molmet.2014.03.007.

Khaksar S, Bigdeli MR. Anti-excitotoxic effects of cannabidiol are partly mediated by enhancement of NCX2 and NCX3 expression in animal model of cerebral ischemia. Eur J Pharmacol. 2016 Nov 14;794:270–9. doi:10.1016/j.ejphar.2016.11.011.

Lipina C, Hundal HS. Modulation of cellular redox homeostasis by the endocannabinoid system. Open Biol. 2016 Apr;6(4):150276. doi:10.1098/rsob.150276.

Ma L, et al. Mitochondrial CB1 receptor is involved in ACEA-induced protective effects on neurons and mitochondrial functions. Sci Rep. 2015 Jul 28;5:12440. doi:10.1038/srep12440.

Mendizabal-Zubiaga J, et al. Cannabinoid CB1 receptors are localized in striated muscle mitochondria and regulate mitochondrial respiration. Front Physiol. 2016 Oct 25;7:476. doi:10.3389/fphys.2016.00476.

Nunn A, Guy G, Bell JD. Endocannabinoids in neuroendopsychology: multiphasic control of mitochondrial function. Philos Trans R Soc Lond B Biol Sci. 2012 Dec 5;367(1607):3342–52. doi:10.1098/rstb.2011.0393.

Penner EA, Buettner H, Mittleman MA. The impact of marijuana use on glucose, insulin, and insulin resistance among US adults. Am J Med. 2013 Jul;126(7):583–9. doi:10.1016/j.amjmed.2013.03.002.

Ryan D, et al. Cannabidiol targets mitochondria to regulate intracellular Ca2+ levels. J Neurosci. 2009 Feb 18;29(7):2053–63. doi:10.1523/JNEUROSCI.4212-08.2009.

Exercise and Physical Activity

Alf D, Schmidt ME, Siebrecht SC. Ubiquinol supplementation enhances peak power production in trained athletes: a double-blind, placebo controlled study. J Int Soc Sports Nutr. 2013 Apr 29;10(1):24. Epub. doi:10.1186/1550-2783-10-24.

Alzheimer's Association International Conference (AAIC); 2012 Jul 14–19; Vancouver, BC. Alzheimers Dement. Abstract F1-03-01.

Alzheimer's Association International Conference (AAIC); 2012 Jul 14–19; Vancouver, BC. Alzheimers Dement. Abstracts FI-03-02.

Alzheimer's Association International Conference (AAIC); 2012 Jul 14–19; Vancouver, BC. Alzheimers Dement. Abstracts P1-109.

Alzheimer's Association International Conference (AAIC); 2012 Jul 14–19; Vancouver, BC. Alzheimers Dement. Abstracts P1-121.

Barrès R, et al. Acute exercise remodels promoter methylation in human skeletal muscle. Cell Metab. 2012 Mar 7;15(3):405–11. doi:10.1016/j.cmet.2012.01.001.

Bergeron R, et al. Chronic activation of AMP kinase results in NRF-1 activation and mitochondrial biogenesis. Am J Physiol Endocrinol Metab. 2001 Dec;281(6):E1340–E1346.

Brown WJ, Pavey T, Bauman AE. Comparing population attributable risks for heart disease across the adult lifespan in women. Br J Sports Med. 2015 Jul 29. Epub 2014 May 8. doi:10.1136/bjsports-2015-095213.

Campbell P, et al. Associations of recreational physical activity and leisure time spent sitting with colorectal cancer survival. J Clin Oncol. 2013 Mar 1;31(7):876–85. doi:10.1200/JCO.2012.45.9735.

Clanton TL. Hypoxia-induced reactive oxygen species formation in skeletal muscle. J Appl Physiol (1985). 2007 Jun;102(6):2379–88. doi:10.1152/japplphysiol.01298.2006.

Díaz-Castro J, et al. Coenzyme Q(10) supplementation ameliorates inflammatory signaling and oxidative stress associated with strenuous exercise. Eur J Nutr. 2012 Oct;51(7):791–9. Epub 2011 Oct 12. doi:10.1007/s00394-011-0257-5.

Erickson KI, et al. Exercise training increases size of hippocampus and improves memory. Proc Natl Acad Sci U S A. 2011 Feb 15;108(7):3017–22. Epub 2011 Jan 31. doi:10.1073/pnas .1015950108.

Gioscia-Ryan RA, et al. Voluntary aerobic exercise increases arterial resilience and mitochondrial health with aging in mice. Aging (Albany NY). 2016 Nov 22;8(11):2897–2914. doi:10.18632/aging.101099.

Gram M, Dahl R, Dela F. Physical inactivity and muscle oxidative capacity in humans. Eur J Sport Sci. 2014;14(4):376–83. Epub 2013 Aug 1. doi:10.1080/17461391.2013.823466.

Greggio C, et al. Enhanced respiratory chain supercomplex formation in response to exercise in human skeletal muscle. Cell Metab. 2017 Feb 7;25(2):301–11. Epub 2016 Dec 1. doi:10.1016/j.cmet.2016.11.004.

Hood DA. Contractile activity-induced mitochondrial biogenesis in skeletal muscle [invited review]. J Appl Physiol (1985). 2001 Mar;90(3):1137–57.

Johnson ML, et al. Chronically endurance-trained individuals preserve skeletal muscle mitochondrial gene expression with age but differences within age groups remain. Physiol Rep. 2014 Dec;2(12):e12239. Epub 2014 Dec 18. doi:10.14814/phy2.12239.

Kang C, et al. Exercise training attenuates aging-associated mitochondrial dysfunction in rat skeletal muscle: role of PGC-1α. Exp Gerontol. 2013 Nov;48(11):1343–50. Epub 2013 Aug 29. doi:10.1016/j.exger.2013.08.004.

Koltai E, et al. Age-associated declines in mitochondrial biogenesis and protein quality control factors are minimized by exercise training. Am J Physiol Regul Integr Comp Physiol. 2012 Jul 15;303(2):R127–R134. Epub 2012 May 9. doi:10.1152/ajpregu.00337.2011.

Konopka AR, et al. Markers of human skeletal muscle mitochondrial biogenesis and quality control: effects of age and aerobic exercise training. J Gerontol A Biol Sci Med Sci. 2014 Apr;69(4):371–8. Epub 2013 Jul 20. doi:10.1093/gerona/glt107.

Konopka AR, et al. Defects in mitochondrial efficiency and H_2O_2 emissions in obese women are restored to a lean phenotype with aerobic exercise training. Diabetes. 2015 Jun;64(6):2104–15. doi:10.2337/db14-1701.

Lawson EC, et al. Aerobic exercise protects retinal function and structure from light-induced retinal degeneration. J Neurosci. 2014 Feb 12;34(7):2406–12. doi:10.1523/JNEUROSCI .2062-13.2014.

Liu CC, et al. Lycopene supplementation attenuated xanthine oxidase and myeloperoxidase activities in skeletal muscle tissues of rats after exhaustive exercise. Br J Nutr. 2005;94: 595–601. doi:10.1079/BJN20051541.

Marcelino TB, et al. Evidences that maternal swimming exercise improves antioxidant defenses and induces mitochondrial biogenesis in brain of young Wistar rats. Neuroscience. 2013 Aug 29;246:28–39. Epub 2013 Apr 29. doi:10.1016/j.neuroscience.2013.04.043.

Melov S, et al. Resistance exercise reverses aging in human skeletal muscle. PLoS One. 2007 May 23;2(5):e465. doi:10.1371/journal.pone.0000465.

Menshikova EV, et al. Effects of exercise on mitochondrial content and function in aging human skeletal muscle. J Gerontol A Biol Sci Med Sci. 2006 Jun;61(6):534–40.

Nikolaidis MG, Jamurtas AZ. Blood as a reactive species generator and redox status regulator during exercise. Arch Biochem Biophys. 2009 Oct 15;490(2):77–84. doi:10.1016/j.abb.2009.08.015.

Powers SK, Jackson MJ. Exercise-induced oxidative stress: cellular mechanisms and impact on muscle force production. Physiol Rev. 2008 Oct;88(4):1243–76. doi:10.1152/physrev.00031.2007.

Reichhold S, et al. Endurance exercise and DNA stability: is there a link to duration and intensity? Mutat Res. 2009 Jul–Aug;682(1):28–38. doi:10.1016/j.mrrev.2009.02.002.

Richardson RS, et al. Myoglobin O2 desaturation during exercise: evidence of limited O2 transport. J Clin Invest. 1995 Oct;96(4):1916–26. doi:10.1172/JCI118237.

Safdar A, et al. Exercise increases mitochondrial PGC-1alpha content and promotes nuclear-mitochondrial cross-talk to coordinate mitochondrial biogenesis. J Biol Chem. 2011 Mar 25;286(12):10605–17. Epub 2011 Jan 18. doi:10.1074/jbc.M110.211466.

Schnohr P, et al. Longevity in male and female joggers: the Copenhagen city heart study. Am J Epidemiol. 2013 Apr 1;177(7):683–9. Epub 2013 Feb 28. doi:10.1093/aje/kws301.

Siddiqui NI, Nessa A, Hossain MA. Regular physical exercise: way to healthy life. Mymensingh Med J. 2010 Jan;19(1):154–8.

Steiner JL, et al. Exercise training increases mitochondrial biogenesis in the brain. J Appl Physiol (1985). 2011 Oct;111(4):1066–71. Epub 2011 Aug 4. doi:10.1152/japplphysiol.00343.2011.

Suzuki K, et al. Circulating cytokines and hormones with immunosuppressive but neutrophil-priming potentials rise after endurance exercise in humans. Eur J Appl Physiol. 2000 Jan;81:281–7.

Toledo FG, et al. Effects of physical activity and weight loss on skeletal muscle mitochondria and relationship with glucose control in type 2 diabetes. Diabetes. 2007 Aug;56(8):2142–7. Epub 2007 May 29. doi:10.2337/db07-0141.

Urso ML, Clarkson PM. Oxidative stress, exercise, and antioxidant supplementation. Toxicology. 2003;189(1–2):41–54. doi:10.1016/S0300-483X(03)00151-3.

Yuki A, et al. Relationship between physical activity and brain atrophy progression. Med Sci Sports Exerc. 2012 Dec;44(12):2362–8. doi:10.1249/MSS.0b013e3182667d1d.

Zong H, et al. AMP kinase is required for mitochondrial biogenesis in skeletal muscle in response to chronic energy deprivation. Proc Natl Acad Sci U S A. 2002 Dec 10;99(25):15983–7. Epub 2002 Nov 20. doi:10.1073/pnas.252625599.

FIGURE CREDITS

Figure 1.1: Kelvinsong, from Wikimedia Commons http://en.wikipedia.org/wiki
/File:Mitochondrion_structure.svg

Figures 1.2 and 1.3: Erin Ford

Figure 1.4: Fvasconcellos, from Wikimedia Commons http://en.wikipedia.org/wiki
/File:Complex_I.svg

Figure 1.5: Fvasconcellos, from Wikimedia Commons http://en.wikipedia.org/wiki
/File:Complex_II.svg

Figure 1.6: Fvasconcellos, from Wikimedia Commons http://en.wikipedia.org/wiki
/File:Complex_III_reaction.svg

Figure 1.7: Fvasconcellos, from Wikimedia Commons http://en.wikipedia.org/wiki
/File:Complex_IV.svg

Figure 1.8: Alex.X, from Wikipedia http://en.wikipedia.org/wiki/File:Atp_synthase.PNG

Figure 1.9: Fvasconcellos, from Wikimedia Commons http://en.wikipedia.org/wiki
/File:Mitochondrial_electron_transport_ chain%E2%80%94Etc4.svg

Figure 2.1: Erin Ford

Figure 3.1: Slagt, from Wikipedia http://en.wikipedia.org/wiki/File:Acyl-CoA_from_cytosol
_to_the_mitochondrial_ matrix.svg

INDEX

ABOUT THE AUTHOR

Lee Know, ND, is a licensed naturopathic doctor based in Canada, and the recipient of several awards. Known by his peers to be a strategic and forward-thinking entrepreneur and physician, he has held positions as medical advisor, scientific evaluator, and director of research and development for major organizations. Besides managing Scientific Affairs for his own company, he also currently serves as a consultant to the natural-health-products and dietary-supplements industries, and serves on the editorial advisory board for *Alive* magazine, Canada's most-read natural health magazine. He calls the Greater Toronto Area home, where he lives with his common-law partner and their two sons, and has a particular interest in promoting natural health and environmental stewardship.

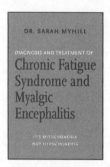
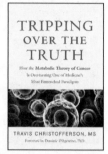